Vibrations Waves and Heat Energy

Calculations in Physics for High Schools and Colleges

By

Kingsley Augustine

TABLE OF CONTENTS

CHAPTER 1 SIMPLE HARMONIC MOTION 3
CHAPTER 2 ENERGY IN SIMPLE HARMONIC MOTION 21
CHAPTER 3 CIRCULAR MOTION 30
CHAPTER 4 WAVE MOTION 38
CHAPTER 5 ECHOES 53
CHAPTER 6 BEAT 61
CHAPTER 7 VIBRATION OF AIR COLUMN IN PIPES 64
CHAPTER 8 MODES OF VIBRATION OF A STRETCHED STRING 74
CHAPTER 9 CHARACTERISTICS OF SOUND – THE PITCH 87
CHAPTER 10 DOPPLER EFFECTS IN SOUND 91
CHAPTER 11 LINEAR EXPANSIVITY 101
CHAPTER 12 AREA EXPANSIVITY 109
CHAPTER 13 VOLUME EXPANSIVITY 116
CHAPTER 14 REAL AND APPARENT CUBIC EXPANSIVITY 124
CHAPTER 15 MEASUREMENT OF TEMPERATURE 1
CHAPTER 16 HEAT ENERGY – HEAT CAPACITY AND SPECIFIC HEAT CAPACITY 11
CHAPTER 17 CHANGE OF STATE – LATENT HEAT AND SPECIFIC LATENT HEAT 22
CHAPTER 18 RELATIVE HUMIDITY 29
CHAPTER 19 BOYLE'S LAW 34
CHAPTER 20 CHARLES' LAW 48
CHAPTER 21 PRESSURE LAW 56
CHAPTER 22 GENERAL GAS LAW 64
ANSWERS TO EXERCISES 73

CHAPTER 1
SIMPLE HARMONIC MOTION

Simple harmonic motion (SHM) is defined as the motion of a body whose acceleration is directly proportional to the displacement from a fixed point and is always directed towards that fixed point.

Three common examples of bodies in simple harmonic motion are:
1. A swinging pendulum bob
2. A displaced mass hanging from a spiral spring
3. A loaded test tube depressed and released in a liquid

Velocity and Acceleration of Simple Harmonic Motion (SHM)

The maximum velocity obtained by a body in SHM is given by:

$$v = \omega A$$

where ω = angular velocity in radians/second (rad/sec), and A = maximum displacement of the body (i.e. the amplitude).

However, at a point which is at a distance of x, from the center of motion (equilibrium or fixed point), the velocity is given by:

$$v = \omega\sqrt{A^2 - x^2}$$

The velocity is zero at the ends of motion, and maximum at the center of motion.

The acceleration, a, for a body performing SHM is given by:

$$a = \omega^2 x$$

where x is the distance of the body from the center of the motion.
The acceleration is zero at the center of motion and maximum at the ends of motion.

It should be noted that this acceleration is negative.

Period and Frequency of SHM

1. **Period**: This is the time taken to complete one cycle or oscillation or vibration. In terms of number of oscillation/cycle, and time taken to complete the oscillation, period, T, is given by:

$$T = \frac{\text{Time taken}}{\text{Number of oscillations /cycles}}$$

2. **Frequency**: This is the number of cycles or oscillations completed in one second. In terms of number of oscillation/cycle, and time taken to complete the oscillation, frequency, f, is given by:

$$f = \frac{\text{Number of oscillations /cycles}}{\text{time taken}}$$

The unit of period is second while the unit of frequency is Hertz (Hz).

Generally, for any object performing SHM, the following formulas apply:

- Angular velocity, $\omega = 2\pi f$ or $\omega = \frac{2\pi}{T}$

- Period, $T = \frac{2\pi}{\omega}$ or $T = \frac{1}{f}$

- Frequency, $f = \frac{\omega}{2\pi}$ or $f = \frac{1}{T}$

Period of Simple Pendulum

Another formula that can be used to calculate the period of a simple pendulum is given by:

$$T = 2\pi\sqrt{\frac{l}{g}}$$

where l is the length of the pendulum.

The equation above shows that the angular velocity for a simple pendulum is given by:

$\omega = \sqrt{\frac{g}{l}}$ (when $T = \frac{2\pi}{\omega}$ is compared with $T = 2\pi\sqrt{\frac{l}{g}}$)

When a simple pendulum of length l_1 and period T_1 is compared to another pendulum of length l_2 and period T_2, then the relation between them is given by:

$$\frac{T_1}{T_2} = \sqrt{\frac{l_1}{l_2}}$$

Or, $\left(\frac{T_1}{T_2}\right)^2 = \frac{l_1}{l_2}$

Period of a Spiral Spring

A formula that can be used to calculate the period of a loaded spiral spring in SHM

is:
$$T = 2\pi\sqrt{\frac{m}{k}} \quad \text{or} \quad T = 2\pi\sqrt{\frac{e}{g}}$$

where m is the mass attached to the spiral spring, k is the elastic constant of the spring, and e is the extension produced by the mass attached to the spring.

The equation above shows that the angular velocity for a spiral spring is given by:

$$\omega = \sqrt{\frac{k}{m}} \quad \text{or} \quad \omega = \sqrt{\frac{g}{e}} \quad (\text{when } T = \frac{2\pi}{\omega} \text{ is compared with } T = 2\pi\sqrt{\frac{m}{k}} \text{ and } T = 2\pi\sqrt{\frac{e}{g}})$$

When a spiral spring has mass, m_1, attached to it and it has period T_1, and it is compared to a similar spiral spring with mass, m_2, attached to it and has period T_2, then the relation between them is given by:

$$\frac{T_1}{T_2} = \sqrt{\frac{m_1}{m_2}}$$

Or, $\left(\frac{T_1}{T_2}\right)^2 = \frac{m_1}{m_2}$

Similarly, when the extension, e_1, on a spiral spring of period T_1, is compared to a similar spiral spring with extension, e_2, and period, T_2, then the relation between them is given by:

$$\frac{T_1}{T_2} = \sqrt{\frac{e_1}{e_2}}$$

Or, $\left(\frac{T_1}{T_2}\right)^2 = \frac{e_1}{e_2}$

Period of a Loaded Test Tube in a Liquid

For a loaded test tube depressed in a liquid and allowed to perform SHM, if its mass is m, cross section area is A, and density of the liquid is ρ, then the period is given by:

$$T = 2\pi\sqrt{\frac{m}{A\rho g}}$$

The equation above shows that the angular velocity for a loaded test tube is given by:

$$\omega = \sqrt{\frac{A\rho g}{m}} \quad (\text{when } T = \frac{2\pi}{\omega} \text{ is compared with } T = 2\pi\sqrt{\frac{m}{A\rho g}})$$

When we compare two similar loaded test tubes in the same liquid, then the relationship connecting their periods and masses is given by:

$$\frac{T_1}{T_2} = \sqrt{\frac{m_1}{m_2}}$$

Or, $\left(\frac{T_1}{T_2}\right)^2 = \frac{m_1}{m_2}$

Equations of Simple Harmonic Motion

The generalized equation of the position of a body in simple harmonic motion as a function of time is given by:

$$x = A \cos(\omega t + \phi)$$

where t is the time in seconds, ω is the angular velocity/frequency, A is the amplitude, and ϕ is the phase difference/shift in radians. When the equation above is differentiated, it gives the equation of the velocity of a body in SHM as follows:

$$v = -A\omega \sin(\omega t + \phi)$$

When the equation for velocity is differentiated, it gives the equation for acceletation of a body in SHM as follows:

$$a = -A\omega^2 \cos(\omega t + \phi)$$

Examples

1. A body performing simple harmonic motion has a maximum displacement from the center of motion to be 0.2m. If its angular velocity is 6rad/sec, calculate:
(a) the period
(b) the frequency
(c) the maximum velocity
(d) the acceleration at the center and at the end of motion
(e) the velocity of the body at a point 0.12m from the center of motion
(Take π = 3.142)

Solution

(a) $T = \frac{2\pi}{\omega}$

$= \frac{2 \times 3.142}{6}$

= 1.05sec

(b) $f = \dfrac{\omega}{2\pi}$

$= \dfrac{6}{2 \times 3.142}$

$= 0.95 Hz$

(c) $v = \omega A$

$= 6 \times 0.2$

$v = 1.2 m/s$

(d) The acceleration at the center is zero.
The acceleration at the end of motion is maximum, and is given by:

$a = \omega^2 x$

$= 6^2 \times 0.2$

$= 36 \times 0.2$

$= 7.2 ms^{-2}$

(e) This velocity at a certain point is given by:

$v = \omega\sqrt{A^2 - x^2}$

$= 6 \times \sqrt{0.2^2 - 0.12^2}$

$= 6 \times \sqrt{0.04 - 0.0144}$

$= 6 \times \sqrt{0.0256}$

$= 6 \times 0.16$

$= 0.96 m/s$

2. A simple pendulum of length 60cm oscillates with amplitude of 0.05m. Calculate:
(a) the period of oscillation
(b) the maximum velocity of the motion   (Take g = 10m/s²)

Solution

(a) $T = 2\pi\sqrt{\dfrac{l}{g}}$

$= 2 \times 3.142 \times \sqrt{\dfrac{0.6}{10}}$ (Note that 60cm = $(\dfrac{60}{100})m = 0.6m$)

$= 6.284 \times \sqrt{0.06}$

$= 1.54 sec$

(b) v = ωA

But, ω = $\frac{2\pi}{T}$

$= \frac{2 \times 3.142}{1.54}$

ω = 4.0805 radsec^{-1}

Hence, v = ωA

= 4.0805 x 0.05

= 0.204 ms^{-1}

3. An object performing simple harmonic motion has an angular velocity of 5rad/sec. If the amplitude of the motion is 30cm, calculate the velocity of the object at a point:

(a) 20cm from the equilibrium position

(b) 12cm from the end of motion

Solution

(a) The amplitude, A = 30cm = $\frac{30}{100}$ = 0.3m

$x = \frac{20}{100} = 0.2m$

The velocity at a certain point is given by:

v = ω$\sqrt{A^2 - x^2}$

= 5 x $\sqrt{0.3^2 - 0.2^2}$

= 5 x $\sqrt{0.09 - 0.04}$

= 5 x $\sqrt{0.05}$

= 5 x 0.2236

= 1.118 m/s

(b) In this case, the distance of the object from the center of motion is given by:

x = 0.3 – 0.12   (Note that 12cm = 0.12m)

x = 0.18

Hence the velocity at 0.12m from end of motion (i.e. 0.18m from center) is given by:

$$v = \omega\sqrt{A^2 - x^2}$$
$$= 5 \times \sqrt{0.3^2 - 0.18^2}$$
$$= 5 \times \sqrt{0.09 - 0.0324}$$
$$= 5 \times \sqrt{0.0576}$$
$$= 5 \times 0.24$$
$$= 1.2 \, m/s$$

4. The acceleration of body in simple harmonic motion is 64 times its displacement in meters. Determine the frequency of the motion.

Solution

Displacement is x. Hence, acceleration is $64x$. Let us now substitute each value into the formula for acceleration as follows:

$$a = \omega^2 A$$
$$64x = \omega^2 x$$
$$64 = \omega^2 \quad (x \text{ has cancelled out})$$
$$\omega = \sqrt{64}$$
$$\omega = 8$$

But, $f = \dfrac{\omega}{2\pi}$

$$= \dfrac{8}{2 \times 3.142}$$
$$f = 1.27 Hz$$

5. The period of a simple pendulum P is 3sec. What is the period of a simple pendulum Q which makes 250 vibrations in the time it takes P to make 200 vibrations.

Solution

Recall that: $T = \dfrac{\text{Time taken}}{\text{Number of oscillations/cycles}}$

Substituting values for pendulum P gives:

$3 = \dfrac{t}{200}$ (t is the time taken by pendulum P)

$t = 3 \times 200$

t = 600sec

This duration of 600 seconds taken by P is also the same time taken by Q. Hence, we now substitute appropriate values for Q in order to obtain its period.

$$T = \frac{\text{Time taken}}{\text{Number of oscillations /cycles}}$$

$$= \frac{600}{250}$$

T = 2.4

Therefore, the period of pendulum Q is 2.4 seconds.

6. An object moving with simple harmonic motion has an amplitude of 8cm and a frequency of 60Hz. Calculate:
(a) the period of oscillation
(b) the velocity at the middle and end of oscillation

Solution

(a) The angular velocity is given by:

$\omega = 2\pi f$

= 2 x 3.142 x 60

ω = 377.04rad/sec

The period is given by:

$$T = \frac{2\pi}{\omega}$$

$$= \frac{2 \times 3.142}{377.04}$$

= 0.0167sec

(b) The velocity at the middle of oscillation is the maximum velocity given by:

v = ωA

= 377.04 x 0.08   (Note that 8cm = 0.08m)

= 30.16ms^{-1}

7. The period of a simple pendulum is 4.4sec. When the length of the pendulum is reduced by 1m, the period is 3.8sec. Determine:

(a) the original length of the pendulum
(b) the value of the acceleration due to gravity of the place.

Solution

(a) The formula for comparing the periods of two simple pendulums is given by:

$$\left(\frac{T_1}{T_2}\right)^2 = \frac{l_1}{l_2}$$

The original length is l_1. Hence, the new length is $l_2 = l_1 - 1$ (since original length was reduced by 1m to obtain the new final length). Substituting into the formula above gives:

$$\left(\frac{T_1}{T_2}\right)^2 = \frac{l_1}{l_2}$$

$$\left(\frac{4.4}{3.8}\right)^2 = \frac{l_1}{l_1 - 1}$$

$$1.3407 = \frac{l_1}{l_1 - 1}$$

$$1.3407(l_1 - 1) = l_1$$

$$1.3407 l_1 - 1.3407 = l_1$$

$$1.3407 l_1 - l_1 = 1.3407$$

$$0.3407 l_1 = 1.3407$$

$$l_1 = \frac{1.3407}{0.3407}$$

$$l_1 = 3.94$$

Therefore, the original length of the pendulum is 3.94m.

(b) Let us use the formula for the period to obtain g as follows.

$$T = 2\pi \sqrt{\frac{l}{g}}$$

$4.4 = 2 \times 3.142 \times \sqrt{\frac{3.94}{g}}$ (use the period of the original pendulum)

$$4.4 = 6.284 \sqrt{\frac{3.94}{g}}$$

Square both sides of the equation to obtain:

$4.4^2 = 6.284^2 \left(\frac{3.94}{g}\right)$ (Note that the root sign has gone)

$$19.36 = \frac{155.585}{g}$$

$$g = \frac{155.585}{19.36}$$

$$g = 8.04$$

Therefore the value of the acceleration due to gravity is $8.04 ms^{-2}$.

8. The period of a simple pendulum is 10sec. Calculate its period when its length is tripled.

Solution

The formula for comparing the periods of two simple pendulums is given by:

$$\left(\frac{T_1}{T_2}\right)^2 = \frac{l_1}{l_2}$$

The original length is l_1. Hence, the new length is $l_2 = 3l_1$ (since original length was tripled). Substituting into the formula above gives:

$$\left(\frac{T_1}{T_2}\right)^2 = \frac{l_1}{l_2}$$

$$\left(\frac{10}{T_2}\right)^2 = \frac{l_1}{3l_1}$$

$$\frac{100}{T_2^2} = \frac{1}{3} \quad (l_1 \text{ has cancelled out})$$

$$= 3 \times 100$$

$$T_2 = \sqrt{300}$$

$$T_2 = 17.32$$

Therefore, the period is 17.32 seconds.

This shows that the new period can be obtained by applying the formula:

$$T_2 = \sqrt{factor\ of\ multiplication} \times T_1$$

In our example above: $T_2 = \sqrt{3} \times 10$

$$= 1.732 \times 10$$

$$= 17.32 \text{ (as obtained above)}$$

9. The period of a spiral spring is 5sec when the mass attached to it is 20g. What is its period when this mass is replaced with an 80g mass?

Solution
The formula for comparing the periods of a spiral spring is given by:

$$\left(\frac{T_1}{T_2}\right)^2 = \frac{m_1}{m_2}$$

Substituting values into the formula above gives:

$$\left(\frac{5}{T_2}\right)^2 = \frac{20}{80}$$

$$\frac{25}{T_2^2} = \frac{1}{4}$$

$T_2 = 4 \times 25$

$T_2 = \sqrt{100}$

$T_2 = 10$

Therefore, the period is 10 seconds.

10. A simple pendulum of length 40cm performs simple harmonic motion. Determine the angular velocity of the pendulum. ($g = 10m/s^2$)

Solution
For a simple pendulum, the angular velocity can be obtained as follows:

$\omega = \sqrt{\frac{g}{l}}$

$= \sqrt{\frac{10}{0.4}}$ (Note that 40cm = 0.4m)

$= \sqrt{25}$

$\omega = 5\,rad/sec$

11. A spiral spring has a mass of 20g attached to it, and this extends the spring by 5cm. The spring is pulled down a distance of 10cm and allowed to perform simple harmonic motion. Calculate:
(a) the angular velocity of the motion
(b) the velocity of the spring at the point 8cm from the mean position.
   ($g = 10m/s^2$)

Solution
(a) For a spiral spring, the angular velocity can be obtained as follows:

$$\omega = \sqrt{\frac{g}{e}}$$
$$= \sqrt{\frac{10}{0.05}} \quad \text{(Note that 5cm = 0.05m)}$$
$$= \sqrt{200}$$
$$\omega = 14.14 \text{rad/sec}$$

(b) $v = \omega\sqrt{A^2 - x^2}$

where A = 10cm = 0.1m and x = 8cm = 0.08m. Substituting these values into the equation above gives:

$$v = 14.14\sqrt{0.1^2 - 0.08^2}$$
$$= 14.14\sqrt{0.0036}$$
$$= 14.14 \times 0.06$$
$$v = 0.848 \text{m/s}$$

12. The equation of the displacement in meters, of a body in simple harmonic motion is given by:

$$x = 6 \cos\left(\pi t + \frac{\pi}{3}\right)$$

where t is in seconds. Determine:
(a) the amplitude, frequency and period of motion
(b) the equations of the velocity and acceleration of the body
(c) position, velocity and acceleration of the body at time t = 1 second

Solution

(a) The given equation: $x = 6 \cos\left(\pi t + \frac{\pi}{3}\right)$

The general equation: $x = A \cos(\omega t + \phi)$

Comparing the two equations above shows that the amplitude, A = 6m.

It also shows that the angular velocity, $\omega = \pi$
Hence, $2\pi f = \pi$ (Note that $\omega = 2\pi f$)
Therefore, $f = \frac{\pi}{2\pi}$ (When both sides are divided by 2π)

$f = \frac{1}{2}$ (π cancels out)

f = 0.5Hz

The period is given by:
$$T = \frac{1}{f}$$
$$= \frac{1}{0.5}$$
T = 2 seconds

(b) $x = 6 \cos(\pi t + \frac{\pi}{3})$

$x = A \cos(\omega t + \phi)$

Comparing the equations above shows that: A = 6, $\omega = \pi$, and $\phi = \frac{\pi}{3}$

The general equation for the velocity of a body in SHM is given by:
$$v = -A\omega \sin(\omega t + \phi)$$

Substituting the appropriate values into the equation above gives the equation for the velocity of the body as follows:
$$v = -6\pi \sin(\pi t + \frac{\pi}{3})$$

The general equation for the acceleration of a body in SHM is given by:
$$a = -A\omega^2 \cos(\omega t + \phi)$$

Substituting the appropriate values into the equation above gives the equation for the acceleration of the body as follows:
$$a = -6\pi^2 \cos(\pi t + \frac{\pi}{3})$$

(c) The equation for the position of the body is given by:
$$x = 6 \cos(\pi t + \frac{\pi}{3})$$

At time t = 1 second, the position is obtained by substituting 1 for t in the equation above. This gives:
$$x = 6 \cos(\pi(1) + \frac{\pi}{3})$$
$$= 6 \cos(\pi + \frac{\pi}{3})$$
$$= 6 \cos \frac{4\pi}{3}$$
$$= 6 \times (-0.5) \quad \text{(Note that } \cos \frac{4\pi}{3} = -0.5, \text{ where } \frac{4\pi}{3} \text{ is in radians not degree)}$$
$$x = -3 \text{m}$$

The equation for the velocity of the body is given by:

$$v = -6\pi \sin(\pi t + \frac{\pi}{3})$$

At time t = 1 second, the velocity is obtained by substituting 1 for t in the equation above. This gives:

$$v = -6\pi \sin(\pi(1) + \frac{\pi}{3})$$

$$= -6\pi \sin(\pi + \frac{\pi}{3})$$

$$= -6\pi \sin \frac{4\pi}{3}$$

$$= -6 \times \pi \times (-0.866) \quad \text{(Note that } \sin \frac{4\pi}{3} = -0.866\text{, where } \frac{4\pi}{3} \text{ is in radians)}$$

$$v = 5.196 \times 3.142 \quad \text{(Note that } \pi = 3.142\text{)}$$

$$v = 16.33 m/s$$

The equation for the acceleration of the body is given by:

$$a = -6\pi^2 \cos(\pi t + \frac{\pi}{3})$$

At time t = 1 second, the acceleration is obtained by substituting 1 for t in the equation above. This gives:

$$a = -6\pi^2 \cos(\pi(1) + \frac{\pi}{3})$$

$$= -6\pi^2 \cos(\pi + \frac{\pi}{3})$$

$$= -6\pi^2 \cos \frac{4\pi}{3}$$

$$= -6 \times \pi \times \pi \times (-0.5)$$

$$a = 3 \times 3.142 \times 3.142$$

$$a = 29.62 m/s^2$$

Note that in working with angles that are in radians, your calculator has to be set to radians not degrees. However, the angles in radians in the problem above can be converted to angles in degrees by simply substituting 180 for π, because 3.142 radians = 180 degrees (i.e. π radians = 180°). For example:

$$\cos \frac{4\pi}{3} = \cos(\frac{4 \times 180}{3}) = \cos 240 = -0.5$$

Hence, $\cos \frac{4\pi}{3} = -0.5$ (where $\frac{4\pi}{3}$ is in radians)

Or, cos 240 = –0.5   (where 240 is in degrees)
So, you can decide to work in radians or degrees.

13. The equation of the displacement of a body in simple harmonic motion is given by:

$x = 2 \cos(5t + 1)$

where x is in meters and t is in seconds. Calculate:
(a) the maximum speed
(b) the maximum acceleration
(c) the displacement of the body between time t = 0 and t = 1 second

Solution
(a) The given equation: $x = 2 \cos(5t + 1)$
 The general equation: $x = A \cos(\omega t + \phi)$
Comparing the two equations above shows that the amplitude, A = 2m, while the angular velocity, ω = 5
Hence, the maximum speed is given by:
   v = ωA
     = 5 x 2
   v = 10m/s

(b) The maximum acceleration is given by:
   a = ω^2A
     = 5^2 x 2
   a = 50ms^{-2}

(c) The equation for the position of the body is given by:
   $x = 2 \cos(5t + 1)$
At time t = 0, the position is obtained by substituting 0 for t in the equation above. This gives:
   $x = 2 \cos(5(0) + 1)$
     = 2 cos 1
     = 2 x (0.5403)   (Note that cos 1 = 0.5403, where 1 is in radians not degree)
   x = 1.08m

At time t = 1, the position is obtained by substituting 1 for t in the equation above. This gives:

$x = 2 \cos (5(1) + 1)$

$= 2 \cos 6$

$= 2 \times (0.9602)$

$x = 1.92m$

Therefore, between t = 0 to t = 1 sec, the body has been displace by:

$1.92 - 1.08 = 0.84m$

Exercise 1

1. A body undergoing simple harmonic motion has a maximum displacement from the center of motion as 10cm. If its angular velocity is 2rad/sec, calculate:
(a) the period
(b) the frequency
(c) the maximum velocity
(d) the acceleration at the center and at the end of motion
(e) the velocity of the body at a point 5cm from the center of motion
   (Take π = 3.142)

2. A simple pendulum of length 100cm oscillates with amplitude of 10cm. Calculate:
(a) the period of oscillation
(b) the maximum velocity of the motion   (Take g = 10m/s^2)

3. An object in simple harmonic motion has an angular velocity of 8rad/sec. If the amplitude of the motion is 15cm, calculate the velocity of the object at a point:
(a) 12cm from the equilibrium position
(b) 7cm from the end of motion

4. The acceleration of body in simple harmonic motion is 20 times its displacement in meters. Determine the frequency of the motion.

5. The period of a simple pendulum X is 5sec. What is the period of a simple pendulum Y which makes 50 vibrations in the time it takes X to make 60 vibrations.

6. An object moving with simple harmonic motion has an amplitude of 6cm and a frequency of 50Hz. Calculate:
(a) the period of oscillation
(b) the velocity at the middle and end of oscillation

7. The time it takes a simple pendulum to complete one vibration is 2.8sec. When the length of the pendulum is reduced by 2m, the period is 1.6sec. Determine:
(a) the original length of the pendulum
(b) the value of the acceleration due to gravity of the place.

8. The period of a simple pendulum is 20sec. Calculate its period when its length is doubled.

9. The period of a spiral spring is 8sec when the mass attached to it is 50g. What is its period when this mass is replaced with a 40g mass?

10. A simple pendulum of length 60cm performs simple harmonic motion. Determine the angular velocity of the pendulum. ($g = 10m/s^2$)

11. A spiral spring has a mass of 120g attached to it, and this extends the spring by 10cm. The spring is pulled down a distance of 6cm and allowed to perform simple harmonic motion. Calculate:
(a) the angular velocity of the motion
(b) the velocity of the spring at the point 2cm from the mean position.
    ($g = 10m/s^2$)

12. The equation of the displacement in meters, of a body performing simple harmonic motion is given by:
$$x = 10 \cos\left(2\pi t + \frac{\pi}{4}\right)$$
where t is in seconds. Determine:
(a) the amplitude, frequency and period of motion
(b) the equations of the velocity and acceleration of the body
(c) position, velocity and acceleration of the body at time t = 2 seconds

13. The equation of the displacement of a body in simple harmonic motion is given by:

$$x = 8\cos(4t + 3)$$

where x is in meters and t is in seconds. Calculate:

(a) the maximum speed

(b) the maximum acceleration

(c) the displacement of the body between time $t = 0$ and $t = 0.5$ seconds

14. The period of a simple pendulum is 12sec. Calculate its period when its length is quartered.

15. Two different simple pendulums perform simple harmonic motion. The ratio of their periods is 2 : 1. Find the ratio of the length of the shorter one to the longer one.

CHAPTER 2
ENERGY IN SIMPLE HARMONIC MOTION

If a spiral spring has a mass m attached to it, and it is pulled down a distance of A, and then released to perform simple harmonic motion, then the maximum potential energy of the motion is given by:

$$P.E_{max} = \frac{1}{2}kA^2$$

where A = amplitude, $k = \frac{F}{e}$, and it is the force constant of the spring, while e is the extension produced by the mass attached.

The potential energy of the motion at a point x from the center of motion is given by:

$$P.E = \frac{1}{2}kx^2$$

This simplifies to: $P.E = \frac{1}{2}m\omega^2 x^2$ (since $k = m\omega^2$)

The maximum kinetic energy of the motion is given by:

$$K.E_{max} = \frac{1}{2}mv^2$$

This simplifies to:

$$K.E_{max} = \frac{1}{2}m\omega^2 A^2$$ (since $v = \omega A$)

However, the kinetic energy at a point x from the fixed point is given by:

$$K.E = \frac{1}{2}k(A^2 - x^2) \quad \text{or} \quad K.E = \frac{1}{2}m\omega^2(A^2 - x^2)$$

Potential energy is maximum at the ends of motion and zero at the center of motion while kinetic energy is maximum at the center of motion and zero at the ends of motion. The maximum kinetic energy is equal to the maximum potential energy.

Hence: $P.E_{max} = K.E_{max}$

Note that at any point in the motion of a body in SHM:

   P.E + K.E = Constant = $P.E_{max}$ and $K.E_{max}$

Examples

1. A light spiral spring is loaded with a mass of 0.05kg and it extends by 0.1m. It is

then pulled vertically down by 0.08m and allowed to perform simple harmonic motion. Calculate:

(a) the period of the spring
(b) the maximum potential energy of the motion
(c) the maximum kinetic energy of the motion
   ($g = 10m/s^2$)

Solution

(a) $T = 2\pi\sqrt{\dfrac{e}{g}}$

$= 2 \times 3.142 \times \sqrt{\dfrac{0.1}{10}}$

$= 6.284 \times \sqrt{0.01}$

$= 6.284 \times 0.1$

$T = 0.628$ sec.

(b) $P.E_{max} = \dfrac{1}{2}kA^2$

But, $k = \dfrac{F}{e}$

$= \dfrac{mg}{0.1}$ (Note that F = W = mg, where W is weight)

$= \dfrac{0.05 \times 10}{0.1}$

$k = 5N/m$

Hence, $P.E_{max} = \dfrac{1}{2}kA^2$

$= \dfrac{1}{2} \times 5 \times 0.08^2$ (Note that the amplitude is 0.08m)

$= \dfrac{1 \times 5 \times 0.0064}{2}$

$P.E = 0.016J$

(c) $K.E_{max} = \dfrac{1}{2}m\omega^2 A^2$

But, $\omega = \dfrac{2\pi}{T}$

$= \dfrac{2 \times 3.142}{0.628}$

$\omega = 10 \text{rad/sec}$

Hence, $K.E_{max} = \frac{1}{2}m\omega^2 A^2$

$= \frac{1}{2} \times 0.05 \times 10^2 \times 0.08^2$

$= \frac{1 \times 0.05 \times 100 \times 0.0064}{2}$

$K.E_{max} = 0.016J$

This shows that the maximum kinetic energy is equal to the maximum potential energy.

2. When a load of 20g is attached to a spiral spring, an extension of 0.04m is produced. If the string is pulled vertically down by 18cm and allowed to perform SHM, calculate:

(a) the period of oscillation

(b) the kinetic energy of the motion at 14cm from the center of oscillation.

Solution

(a) Given: $m = \frac{20}{1000} = 0.02\text{kg}$, $A = \frac{18}{100} = 0.18\text{m}$, $e = 0.04\text{m}$.

$T = 2\pi\sqrt{\frac{e}{g}}$

$= 2 \times 3.142 \times \sqrt{\frac{0.04}{10}}$

$= 6.284 \times \sqrt{0.004}$

$= 6.284 \times 0.0632$

$T = 0.397\text{sec}.$

(b) $K.E = \frac{1}{2}k(A^2 - x^2)$

But, $k = \frac{F}{e}$

$= \frac{mg}{0.1}$

$= \frac{0.02 \times 10}{0.04}$

$k = 5N/m$

Hence, K.E $= \frac{1}{2}k(A^2 - x^2)$

$= \frac{1}{2} \times 5(0.18^2 - 0.14^2)$ (Note that $x = 14\text{cm} = 0.14\text{m}$)

$= 2.5(0.0324 - 0.0196)$

K.E $= 0.032\text{J}$

3. A particle of mass 10g performs simple harmonic motion of amplitude 6cm and period 2π seconds. Calculate:
(a) the kinetic energy and potential energy when it is at a distance of 4cm from its equilibrium position.
(b) the maximum kinetic energy of the particle.

Solution

Given: m $= \frac{10}{1000} = 0.01\text{kg}$, A $= \frac{6}{100} = 0.06\text{m}$, $x = 0.04\text{m}$.

(a) K.E $= \frac{1}{2}m\omega^2(A^2 - x^2)$

But, $\omega = \frac{2\pi}{T}$

$= \frac{2\pi}{2\pi}$

$\omega = 1\text{rad/sec}$

Hence, K.E $= \frac{1}{2}m\omega^2(A^2 - x^2)$

$= \frac{1}{2} \times 0.01 \times 1^2(0.06^2 - 0.04^2)$

$= \frac{1 \times 0.01 \times 1 \times 0.002}{2}$

$= 0.00001\text{J}$

The potential energy is given by:

P.E $= \frac{1}{2}m\omega^2 x^2$

$= \frac{1}{2} \times 0.01 \times 1^2 \times 0.04^2$

$= \frac{1 \times 0.01 \times 1 \times 0.0016}{2}$

$= 0.000008\text{J}$

(b) Recall that at any point:
    P.E + K.E = Constant = P.E$_{max}$ and K.E$_{max}$
Hence, P.E + K.E = K.E$_{max}$
    0.00001 + 0.000008 = K.E$_{max}$
    K.E$_{max.}$ = 0.000018J

4. A body of mass 20g performs simple harmonic motion at a frequency of 5Hz. At a distance of 10cm from the mean position, its velocity is 200cm/s. Calculate its:
(a) maximum displacement from the mean position
(b) potential and kinetic energy at this position of 10cm from the mean position
(c) maximum potential and kinetic energy.
    (π = 3.14, g = 10m/s^2)

Solution

Given: m = $\frac{20}{1000}$ = 0.02kg, f = 5Hz, x = 0.1m, v = $\frac{200}{100}$ = 2m/s

(a) Let us first determine the angular velocity as follows:
    ω = 2πf
    = 2 x 3.14 x 5
    ω = 31.4rad/sec

Recall that: v = $\omega\sqrt{A^2 - x^2}$
    2 = 31.4$\sqrt{A^2 - 0.1^2}$

Squaring both sides of the equation gives:
    2^2 = 31.4^2 (A^2 – 0.1^2)
    4 = 985.96(A^2 – 0.01)
    4 = 985.96A^2 – 9.8596
    4 + 9.8596 = 985.96A^2

    A = $\sqrt{\frac{13.8596}{985.96}}$

    A = 0.119

Therefore, the maximum displacement from the mean position is 0.119m

(b) $P.E = \frac{1}{2}m\omega^2 x^2$

$= \frac{1}{2} \times 0.02 \times 31.4^2 \times 0.1^2$

$= \frac{1 \times 0.02 \times 985.96 \times 0.01}{2}$

$P.E = 0.0986J$

$K.E = \frac{1}{2}m\omega^2(A^2 - x^2)$

$= \frac{1}{2} \times 0.02 \times 31.4^2(0.119^2 - 0.1^2)$

$= \frac{1 \times 0.02 \times 985.96 \times 0.004161}{2}$

$K.E = 0.0821J$

(c) Recall that at any point:

$P.E + K.E = $ Constant $= P.E_{max}$ and $K.E_{max}$

Hence, $P.E + K.E = P.E_{max}$ and $K.E_{max}$

$0.0986 + 0.0821 = P.E_{max}$ and $K.E_{max}$

$P.E_{max}$ and $K.E_{max.} = 0.181J$

Therefore, the maximum potential and kinetic energy are 0.181J each.

5. The mass of the bob of a simple pendulum performing simple harmonic motion is 15g. If at the ends of motion, the bob is at a vertical distance of 2cm from the equilibrium point, calculate the maximum kinetic energy of the bob. ($g = 10m/s^2$)

Solution

Given: m = 0.015kg, h = 0.02m and g = $10m/s^2$

The maximum potential energy of the bob is obtained by:

$P.E_{max} = mgh$

$= 0.015 \times 10 \times 0.02$

$= 0.003J$

Recall that: $P.E_{max} = K.E_{max}$

Therefore, the maximum kinetic energy of the bob is 0.003J.

6. A 5kg mass is suspended from the end of a spring and released to perform simple

harmonic motion with amplitude 4cm and a period of 2sec. Find the maximum energy of the mass.

Solution

Let us determine the angular velocity/frequency as follows:

$$\omega = \frac{2\pi}{T}$$
$$= \frac{2 \times 3.142}{2}$$
$$\omega = 3.142 \text{ rads}^{-1}$$

Hence the energy of the mass is given by:

$$E = \frac{1}{2}m\omega^2 A^2$$
$$= \frac{1}{2} \times 5 \times 3.142^2 \times 0.04^2 \quad \text{(Note that 4cm = 0.04m)}$$
$$= \frac{1 \times 5 \times 9.872 \times 0.0016}{2}$$
$$E = 0.0395 \text{ J}$$

Exercise 2

1. A mass of 80g is hung from a spiral spring thereby producing an extension of 0.02m. It is then pulled vertically down by 0.05m and allowed to perform simple harmonic motion. Calculate:

(a) the period of the spring
(b) the maximum potential energy of the motion
(c) the maximum kinetic energy of the motion
   ($g = 10 \text{ m/s}^2$)

2. When a load of 30g is attached to a spiral spring, an extension of 4cm is produced. If the string is pulled vertically down by 8cm and allowed to perform SHM, calculate:

(a) the period of oscillation
(b) the kinetic energy of the motion at 2cm from the center of oscillation.

3. A particle of mass 2g performs simple harmonic motion of amplitude 3cm and period π seconds. Calculate:
(a) the kinetic energy and potential energy when it is at a distance of 1cm from its equilibrium position.
(b) the maximum potential energy of the particle.

4. A body of mass 10g performs simple harmonic motion of frequency of 2Hz. At a distance of 5cm from the mean position, its velocity is 0.8m/s. Calculate its:
(a) maximum displacement from the mean position
(b) potential and kinetic energy at this position of 5cm from the mean position
(c) maximum potential and kinetic energy.
   ($g = 10m/s^2$)

5. The mass of the bob of a simple pendulum performing simple harmonic motion is 9g. If at the ends of motion, the bob is at a vertical distance of 2.4cm from the equilibrium point, calculate the maximum kinetic energy of the bob. ($g = 10m/s^2$)

6. A 200g mass is suspended from the end of a spring and released to perform simple harmonic motion with amplitude 12cm and a period of 16sec. Find the maximum energy of the mass.

7. A particle of mass 6g performs simple harmonic motion of amplitude 5cm and period 10 seconds. Calculate:
(a) the kinetic energy and potential energy when it is at a distance of 2cm from its equilibrium position.
(b) the maximum potential energy of the particle.

8. A spring is loaded with a mass of 0.02kg and it extends by 0.08m. It is then pulled vertically down by 4cm and allowed to perform simple harmonic motion. Calculate:
(a) the period of the spring
(b) the maximum potential energy of the motion
(c) the maximum kinetic energy of the motion
   ($g = 10m/s^2$)

29

CHAPTER 3
CIRCULAR MOTION

When a body moves round a circular path, its motion is described as a circular motion. Its path is a curve which gives an angular distance, θ, measured in radians. Hence the angular velocity, ω, of the body is defined as follows:

$$\omega = \frac{\text{Angle turned through by the body}}{\text{time taken}}$$

$$\omega = \frac{\theta}{t}$$

where θ = angular distance in radians, and t = time in seconds.

The circular distance, s, or length of arc that the body has moved can be obtained by:

$$s = r\theta$$

where r is the radius of the circular path.

Note that 360 degrees which is one cycle is equal to 2π radians or $180° = \pi$ radians. Hence, an angle, α, in degrees can be converted to angle in radians by using the formula below:

$$\theta = \frac{\alpha\pi}{180}$$ (Note that the value of π is usually taken to be 3.142)

The relationship between linear velocity and angular velocity for a body undergoing circular motion is given by:

$$\omega = \frac{v}{r}$$

Or, $v = \omega r$

This shows that circular motion is related to simple harmonic motion (where $v = \omega A$). Generally, the formulas for simple harmonic motion can be used for circular motion.

Hence in circular motion, the linear acceleration of the body is given by:

$$a = \omega^2 r$$

Or, $a = \frac{v^2}{r}$

Linear acceleration is related to angular acceleration by the expression below:

$$\alpha = \frac{a}{r}$$

Or, a = αr

where α is the angular acceleration measured in per square second (s^{-2})

Centripetal Force

When a body moves round a circular path, it experiences a force which is directed towards the center of the circle and keeps the body on the circular path. This force is called centripetal force. It is given by:

$$F = m\omega^2 r \quad \text{(since F = ma and a = } \omega^2 r\text{)}$$

$$\text{Or, } F = \frac{mv^2}{r} \quad \text{(since a is also } \frac{v^2}{r}\text{)}$$

Examples

1. A body moves along a circular path with a uniform angular speed of 0.5rad/sec and at a constant speed of 1m/s. Calculate the acceleration of the body.

Solution

Let us first calculate the radius of the path.

$$v = \omega r$$
$$r = \frac{v}{\omega}$$
$$= \frac{1}{0.5}$$
$$r = 2m$$

Therefore, the acceleration is given by:

$$a = \omega^2 r$$
$$= 0.5^2 \times 2$$
$$= 0.25 \times 2$$
$$a = 0.5 ms^{-2}$$

2. A body rotating on a circular path makes 420r.p.m. What is its frequency?

Solution

420r.p.m means 420 revolutions per minute. This means that the body made 420 cycles (i.e. revolutions) in 60 seconds (i.e. 1 minute).

Recall that: $f = \dfrac{\text{Number of oscillations/cycles}}{\text{time taken}}$

$$f = \frac{420}{60}$$

f = 6Hz

3. An object covers an angular distance of 270° in 8 seconds. Calculate its angular velocity in rad/sec.

Solution

Let us first convert the angle from degrees to radians as follows:

$$\theta = \frac{\alpha \pi}{180}$$ (Note that α = 270°)

$$= \frac{270 \times 3.142}{180}$$

θ = 4.713 radians

Therefore, the angular velocity is obtained as follows:

$$\omega = \frac{\theta}{t}$$

$$= \frac{4.713}{8}$$

= 0.589 radsec^{-1}

4. A particle makes 300rpm on a circular path of radius 60cm. Find:
(a) its period
(b) its angular velocity
(c) its linear velocity
(d) its acceleration
(e) its angular acceleration

Solution

(a) 300rpm means 300 revolutions per minute. This means that the body made 300 cycles (i.e. revolutions) in 60 seconds (i.e. 1 minute).

Recall that: $$T = \frac{time\ taken}{Number\ of\ cycles}$$

$$T = \frac{60}{300}$$

T = 0.2sec

(b) $\omega = \dfrac{2\pi}{T}$

$= \dfrac{2 \times 3.142}{0.2}$

$\omega = 31.42\, \text{radsec}^{-1}$

(c) $v = \omega r$

$= 31.42 \times 0.6$ (60cm = 0.6m)

$= 18.852\, \text{ms}^{-1}$

(d) $a = \omega^2 r$

$= 31.42^2 \times 0.6$

$= 987.2164 \times 0.6$

$= 592.3\, \text{ms}^{-2}$

(e) Angular acceleration is given by:

$\alpha = \dfrac{a}{r}$

$= \dfrac{592.3}{0.6}$

$= 987.2\, \text{s}^{-2}$

5. An object in a circular path covers an angle of 315° in 6sec. If the radius of the circular path is 10cm, calculate:
(a) its angular speed in rad/sec
(b) its linear velocity
(c) its frequency
(d) its maximum acceleration

Solution

(a) Let us first convert the angle from degrees to radians as follows:

$\theta = \dfrac{\alpha \pi}{180}$

$= \dfrac{315 \times 3.142}{180}$

$\theta = 5.4985$ radians

Therefore, the angular speed is obtained as follows:

$$\omega = \frac{\theta}{t}$$
$$= \frac{5.4985}{6}$$
$$= 0.916 \text{ radsec}^{-1}$$

(b) $v = \omega r$
$= 0.916 \times 0.1$ (10cm = 0.1m)
$= 0.0916 \text{ms}^{-1}$

(c) $f = \frac{\omega}{2\pi}$
$= \frac{0.916}{2 \times 3.142}$
$= 0.146 \text{Hz}$

(d) $a = \omega^2 r$
$= 0.916^2 \times 0.1$
$= 0.0839 \text{ms}^{-2}$

6. A body of mass 50g moves a velocity of 2m/s in a circular path of radius 20cm. Calculate the centripetal force acting on the body.

Solution
Given: m = 0.05kg, v = 2m/s, r = 0.2m
The centripetal force on the body is given by:
$$F = \frac{mv^2}{r}$$
$$= \frac{0.05 \times 2^2}{0.2}$$
F = 1N

7. A particle of mass 4g makes 540rpm on a circular path of radius 6cm. Calculate the value of the force that keeps the body on the path.

Solution
Recall that: $f = \frac{\text{Number of cycles}}{\text{time taken}}$

$$= \frac{540}{60}$$

f = 9Hz

$\omega = 2\pi f$
   $= 2 \times 3.142 \times 9$
$\omega = 56.556$ radsec^{-1}

The force required is given by:
   $F = m\omega^2 r$
      $= 0.004 \times 56.556^2 \times 0.06$
   $F = 0.768$ N

8. An object of mass 0.2kg covers an angular distance of 90° in 1.2 seconds. If its path is a circle of radius 0.4m, calculate:
(a) its angular frequency
(b) its linear velocity
(c) the centripetal force acting on the object.

Solution
(a) Let us first convert the angle from degrees to radians as follows:
$$\theta = \frac{\alpha\pi}{180}$$
$$= \frac{90 \times 3.142}{180}$$

$\theta = 1.571$ radians

Therefore, the angular frequency is obtained as follows:
$$\omega = \frac{\theta}{t}$$
$$= \frac{1.571}{1.2}$$
$\omega = 1.309$ radsec^{-1}

(b) $v = \omega r$
   $= 1.309 \times 0.4$
   $= 0.524$ ms^{-1}

(c) The centripetal force is given by:
$$F = m\omega^2 r$$
$$= 0.2 \times 1.309^2 \times 0.4$$
$$F = 0.137 N$$

Or, the centripetal force can also be given by:
$$F = \frac{mv^2}{r}$$
$$= \frac{0.2 \times 0.524^2}{0.4}$$
$$F = 0.137 N \quad \text{(As obtained above)}$$

Exercise 3

1. A body moves along a circular path with a uniform velocity of 1rad/sec and at a constant speed of 2m/s. Calculate the acceleration of the body.

2. A body rotating on a circular path makes 2400r.p.m. What is the frequency of the body?

3. An object covers an angular distance of 60° in 5 seconds. Calculate its angular velocity in rad/sec.

4. A particle makes 360rpm on a circular path of radius 20cm. Find:
(a) its period
(b) its angular velocity
(c) its linear velocity
(d) its acceleration
(e) its angular acceleration

5. An object in a circular path covers an angle of 180° in 4sec. If the radius of the circular path is 80cm, calculate:
(a) its angular speed in rad/sec
(b) its linear velocity

(c) its frequency
(d) its maximum acceleration

6. A body of mass 110g moves at a velocity of 300cm/s in a circular path of radius 8cm. Calculate the centripetal force acting on the body.

7. A particle of mass 14g makes 200rpm on a circular path of radius 2cm. Calculate the value of the force that keeps the body on the path.

8. An object of mass 10g covers an angular distance of 100° in 1.8 seconds. If its path is a circle of radius 0.12m, calculate:
(a) its angular frequency
(b) its linear velocity
(c) the centripetal force acting on the object.

CHAPTER 4
WAVE MOTION

Wave is a disturbance that travels through a medium and transfers energy from one point to another without actually causing a permanent displacement of the medium.

The period of a wave is given by:

$$T = \frac{2\pi}{w}$$ where w = angular velocity. $w = 2\pi f$. This simplifies to also give:

$$T = \frac{1}{f}$$ where f is the frequency of the wave.

The wave velocity of a wave is given by:

$$v = \lambda/T$$ this simplifies to:

$$v = f\lambda \quad \text{(Since } T = \frac{1}{f}\text{)}$$

This equation is applicable to all waves.

A wave can also be represented mathematically as follows:

$$y = A \sin\phi \quad \text{where } \phi = \text{phase angle}$$

But $\phi = \omega(t - \frac{x}{v})$ where ω = angular velocity, t = time, x = horizontal distance, v = velocity and y = vertical distance

$$\therefore \quad y = A \sin \omega(t - \frac{x}{v})$$

Since $\omega = 2\pi f$, the equation above can be expressed as follows:

$$y = A \sin 2\pi f (t - \frac{x}{v})$$

Or $y = A \sin(2\pi ft - \frac{2\pi fx}{v})$

Or $y = A \sin(2\pi ft - 2\pi x/\lambda)$ (Since $\frac{f}{v} = 1/\lambda$ from $v = f\lambda$)

The expression $2\pi/\lambda$ in the equation above is called the wave number, k.

The two wave equations that will be used in the solved examples below are:

$$y = A \sin(2\pi ft - \frac{2\pi fx}{v}) \text{ and}$$

$$y = A \sin(2\pi ft - 2\pi x/\lambda)$$

Examples

1. On a graphical representation of a wave the height of each curve above the *x* axis is 8m, while the width of each curve along the *x* axis is 5m.

a. Find the amplitude and wavelength of the motion

b. Calculate the period of the wave if it has an angular velocity of 6.4rad/sec

c. Calculate the velocity of the wave

Solution

a. The amplitude is the maximum vertical height of the wave. So, amplitude A, is 8m

The wavelength is the width of two curves. This width is also equal to the distance between successive wave crest. So, wavelength, λ = 2 x 5 = 10m

b. w = 6.4rad/sec. But period, T = $\frac{2\pi}{w}$

∴ T = $\frac{2\pi}{6.4} = \frac{2 \times 3.142}{6.4}$

= $\frac{6.284}{6.4}$ = 0.982

The period is 0.982sec.

c. $v = \lambda/T$

$= \dfrac{10}{0.982} = 10.2$

The velocity of the wave is 10.2ms^{-1}

2. Determine the wavelength of a wave motion that travels a distance of 6m which is made up of five curves.

Solution

The width of each of the bell shape curve of a wave is equal to half a wavelength.

∴ Since 1 curve = $\lambda/2$, then 5 curves = $5 \times \lambda/2 = 5\lambda/2$

These 5 curves are 6m wide. This means that:

$5\lambda/2 = 6$

∴ $5\lambda = 6 \times 2$

$\lambda = \dfrac{12}{5} = 2.4$

The wavelength of the wave is 2.4m.

3. A wave having a frequency of 25Hz travels at a velocity of 40m/s. Calculate the wavelength of the motion.

Solution

$v = f\lambda$

∴ $\lambda = \dfrac{v}{f}$

$= \dfrac{40}{25} = 1.6$

The wavelength is 1.6m

4. A wave travels a distance of 42m in 7sec, and the distance between successive wave crests of the wave is 1.2m. Calculate the frequency of the wave.

Solution

$$\text{Velocity} = \frac{Distance}{Time}$$

$$\therefore \quad v = \frac{42}{7} = 6 \text{m/s}$$

Also, $v = f\lambda$

So, $f = v/\lambda$

$$= \frac{6}{1.2} = 5$$

The frequency of the wave is 5Hz.

5. The distance between successive troughs of a wave of frequency 20Hz is 40cm. Calculate the time taken by the wave to cover a distance of 26m.

Solution

Wavelength, λ, is the distance between successive troughs.

$$\therefore \quad \lambda = 40\text{cm} = \left(\frac{40}{100}\right)\text{m} = 0.4\text{m}, \text{ and } f = 20\text{Hz}$$

So, $v = f\lambda$

$$= 20 \times 0.4 = 8 \text{ms}^{-1}$$

But, $v = \frac{Distance}{Time}$

$$\therefore \quad 8 = \frac{26}{Time}$$

Time = $\frac{26}{8}$ = 3.25

The time taken by the wave is 3.25sec.

6. The displacement, y, of a wave travelling in the negative x direction is represented by, y = 5sin12π (t + $\frac{x}{30}$), where x and y are in metres and t in seconds.

a. What is the amplitude of the wave?

b. Determine the frequency of the wave

c. Determine the velocity of the wave

d. What is the wavelength of the wave?

e. Find the wave number of the wave.

Solutions

a. The general equation of a wave is: y = A sin (2πft - $\frac{2\pi fx}{v}$)

The equation of the wave in the question is: y = 5sin12π (t + $\frac{x}{30}$)

Comparing the two equations shows that the amplitude A = 5

∴ The amplitude of the wave is 5m.

b. y = 5sin12π (t + $\frac{x}{30}$). When 12π is used to expand the bracket, the equation becomes:

y = 5sin (12πt + $\frac{12\pi x}{30}$). The general wave equation is: y = A sin (2πft - $\frac{2\pi fx}{v}$)

Comparing these two equations shows that the first terms in both brackets which contain the frequency, f, (i.e. what we want to find) and the time, t, are equal. Note that only the general wave equation will contain what we want to calculate, i.e. the frequency, f. Both terms contain the time, t. Equating them gives:

$2\pi ft = 12\pi t$. Cancelling out the t and π gives:

$2f = 12$

∴ $f = \dfrac{12}{2} = 6$

The frequency of the wave is 6Hz.

c. The expanded equation from the question and the general wave equation are:

$y = 5\sin(12\pi t + \dfrac{12\pi x}{30})$ and, $y = A\sin(2\pi ft - \dfrac{2\pi fx}{v})$

In order to determine the velocity, compare the two equations and equate the equal terms based on what we want to calculate. The terms are those that contain x, and the velocity, v, (i.e. what we want to find).

So, $\dfrac{12\pi x}{30} = \dfrac{2\pi fx}{v}$. Cancelling out the x and π gives:

$\dfrac{12}{30} = \dfrac{2f}{v}$

$12v = 30 \times 2f$

$12v = 60f$

$v = \dfrac{60f}{12}$

$= \dfrac{60 \times 6}{12}$ (Since f = 6 from solution b. above)

$= \dfrac{360}{12} = 30$

So, the velocity of the wave is 30ms^{-1}

d. The general wave equation that contains wavelength is: $y = A\sin(2\pi ft - 2\pi x/\lambda)$

Also, from the question, $y = 5\sin(12\pi t + \frac{12\pi x}{30})$

In order to determine the wavelength equate the terms that contain the wavelength, λ, along with the common distance x.

So, $\frac{2\pi x}{\lambda} = \frac{12\pi x}{30}$. Cancelling out the x and π gives:

$$\frac{2}{\lambda} = \frac{12}{30}$$

$$12\lambda = 2 \times 30$$

$$\lambda = \frac{60}{12} = 5$$

The wavelength is 5m.

e. The wave number, k, is given by:

$$k = \frac{2\pi}{\lambda}.$$

$$= \frac{2 \times 3.142}{5} = \frac{6.284}{5} = 1.26$$

The wave number is 1.26m^{-1}

7. The frequency of a travelling wave is 10Hz. Starting from the origin, its first crest is at $x = 2m$.

   a. What is the wavelength of the wave?
   b. What time does it take the wave to attain its first crest?
   c. Determine the velocity of the wave
   d. If the amplitude of the wave is 5m, what is the equation of the wave?

   Solution

a. The first crest is half of a semi circle, which is also a quarter of a circle. A cycle which gives a wavelength is equal to four quarters of a circle. From the question above, a quarter of a circle (i.e. first crest) is 2m long. Hence a wavelength which is four quarters is given by:

$\lambda = 4 \times 2$

$\lambda = 8m$

The wavelength is 8m

b. The time taken to travel one wavelength is equal to the period, and the period is given by: $T = \dfrac{1}{f}$

$= \dfrac{1}{10}$

$T = 0.1 sec$

∴ It takes 0.1sec to travel a distance of λ, i.e. 8m.

But, the first crest is attained at a quarter of one wavelength, i.e. $\lambda/4$

Similarly, it will also take a quarter of a period to attain the first crest. This is given by:

Time taken $= \dfrac{T}{4}$

$= \dfrac{0.1}{4}$

$= 0.025$

Hence, it takes 0.025sec for the wave to attain its first crest.

c. The velocity of the wave is given by:

$v = f\lambda$

$= 30 \times 8 = 240$

The velocity of the wave is 240m/s

d. The general equation of a wave is given by:

$$y = A \sin(2\pi ft - 2\pi x/\lambda)$$

Factorizing the term in the bracket gives:

$$y = A \sin 2\pi (ft - x/\lambda)$$

Substituting A = 5, f = 10 and λ = 8, gives:

$$\therefore \quad y = 5\sin 2\pi \left(10t - \frac{x}{8}\right)$$

It can also be expressed by substituting the actual value of π as follows:

$$y = 5\sin 2\pi \left(10t - \frac{x}{3}\right)$$

$$y = 5\sin 2 \times 3.142 \left(10t - \frac{x}{3}\right)$$

$$y = 5\sin 6.284 \left(10t - \frac{x}{3}\right)$$

$$y = 5\sin \left(6.284 \times 10t - \frac{6.284x}{3}\right)$$

$$\therefore \quad y = 5\sin(62.8t - 2.1x)$$

8. If y = 2sin (3x − 4t), where x and y are in metres and t in seconds represent a wave motion, Determine the:

a. amplitude

b. frequency

c. period

d. velocity

e. angular velocity

f. wavelength

g. wave number of the wave.

Solutions

a. The general equation of a wave is: $y = A \sin(2\pi f t - \frac{2\pi f x}{v})$

The equation of the wave in the question is: $y = 2\sin(3x - 4t)$

Comparing the two equations shows that the amplitude A = 2

∴ The amplitude of the wave is 2m.

b. $y = 2\sin(3x - 4t)$. The general wave equation is: $y = A \sin(2\pi f t - \frac{2\pi f x}{v})$

Comparing these two equations shows that we equate the terms in both brackets which contain the time t and the frequency, f, (i.e. what we want to find). Note that only the general wave equation will contain what we want to calculate, i.e. the frequency, f. Equating the terms gives:

2πft = 4t. (Note that the negative sign should be ignored).

Cancelling out the t gives:

2πf = 4

∴ $f = \frac{4}{2\pi} = \frac{4}{2 \times 3.142} = \frac{4}{6.284} = 0.64$

The frequency of the wave is 0.64Hz.

c. The period is given by:

$T = \frac{1}{f}$

$= \frac{1}{0.64} = 1.56$

The period is 1.56sec.

d. $y = 2\sin(3x - 4t)$ and $y = A\sin(2\pi ft - \frac{2\pi fx}{v})$

In order to determine the velocity, compare the two equations and equate the equal terms based on what we want to calculate. The terms are those that contain x in the two brackets, and the velocity, v, (i.e. what we want to find) in only one bracket.

So, $3x = \frac{2\pi fx}{v}$. Cancelling out the x gives:

$$3 = \frac{2\pi f}{v}$$

$$3v = 2\pi f$$

$$\therefore v = \frac{2\pi f}{3}$$

$$= \frac{2 \times 3.142 \times 0.64}{3} \quad \text{(Since f = 0.64 from solution b. above)}$$

$$= \frac{4.022}{3} = 1.34$$

The velocity of the wave is 1.34ms^{-1}

e. The angular velocity, w, is given by:

$$\omega = 2\pi f$$

$$= 2 \times 3.142 \times 0.64$$

$$= 4.02 \text{rad/sec}$$

f. The general wave equation that contains wavelength is: $y = A\sin(2\pi ft - 2\pi x/\lambda)$

Also, from the question, $y = 2\sin(3x - 4t)$

In order to determine the wavelength equate the terms that contain the wavelength, λ, (i.e. what we want to find) along with the common distance x.

So, $2\pi x/\lambda = 3x$. Cancelling out the x gives:

$2\pi/\lambda = 3$

$3\lambda = 2\pi$

$$\lambda = \frac{2\pi}{3} = \frac{2 \times 3.142}{3} = \frac{6.284}{3} = 2.09$$

The wavelength is 2.09m.

g. The wave number, k, can also be obtained by comparing the wave equations as follows:

$y = 2\sin(3x - 4t)$ and $y = A\sin(2\pi ft - 2\pi x/\lambda)$

Since $k = 2\pi/\lambda$, then by comparison,

$2\pi x/\lambda = 3x$

∴ $kx = 3x$ (By substituting $2\pi/\lambda$ for k)

Cancelling out x shows that:

$k = 3$

The wave number is 3m^{-1}

9. Waves whose crests are 40cm apart made a cork floating on water in a ripple tank to rise and fall through a total range of 5cm once every 2 seconds. Determine the:

a. amplitude

b. frequency

c. velocity of the wave.

Solutions

a. The range of 5cm is the vertical distance between the crest and the trough. It is twice the amplitude.

$$\therefore \text{ Amplitude} = \frac{1}{2}(\text{range}) = \frac{1}{2} \times 5 = 2.5\text{cm}$$

b. The period is the time taken to rise from the starting horizontal level up to the crest and then to fall to the trough, and finally to the starting level. This took a total time of 2 seconds. It is also the time taken to cover a distance of one wavelength. So, the period, T, is 2 seconds.

But, frequency, f is given by:

$$f = \frac{1}{T}$$

$$= \frac{1}{2} = 0.5$$

The frequency is 0.5Hz

c. The velocity is given by:

$$v = f\lambda$$

But, $\lambda = 40\text{cm} = (\frac{40}{100})\text{m} = 0.4\text{m}$

$$\therefore v = f\lambda$$

$$= 0.5 \times 0.4 = 0.2$$

The velocity of the wave is 0.2ms^{-1}

Exercise 4

1. On a graphical representation of a wave the height of each curve above the x axis is 200cm, while the width of each curve along the x axis is 80cm.

    a. Find the amplitude and wavelength of the motion

    b. Calculate the period of the wave if it has an angular velocity of 5rad/sec

    c. Calculate the velocity of the wave

2. Determine the wavelength of a wave motion that travels a distance of 16m along the x-axis which is made up of ten curves.

3. A wave having a frequency of 40Hz travels at a velocity of 15m/s. Calculate the wavelength of the motion.

4. A wave travels a distance of 300cm in 3.25sec, and the distance between successive wave crests of the wave is 2m. Calculate the frequency of the wave.

5. The distance between successive troughs of a wave of frequency 50Hz is 6m. Calculate the time taken by the wave to cover a distance of 16m.

6. The displacement, y, of a wave travelling in the negative x direction is represented by, $y = 10\sin 8\pi (t + \frac{x}{50})$, where x and y are in metres and t in seconds.

    a. What is the amplitude of the wave?

    b. Determine the frequency of the wave

    c. Determine the velocity of the wave

    d. What is the wavelength of the wave?

    e. Find the wave number of the wave.

7. The frequency of a travelling wave is 25Hz. Its first crest is at $x = 5m$.

    a. What is the wavelength of the wave?

    b. What time does it take the wave to attain its first crest?

    c. Determine the velocity of the wave

d. If the amplitude of the wave is 8m, what is the equation of the wave?

8. If y = 4sin (5x − 2t), where x and y are in metres and t in seconds represent a wave motion, Determine the:

a. amplitude

b. frequency

c. period

d. velocity

e. angular velocity

f. wavelength

g. wave number of the wave.

9. Waves whose crests are 1.2m apart made a cork floating on water in a ripple tank to rise and fall through a total range of 0.14m once every 5 seconds. Determine the:

a. amplitude

b. frequency

c. velocity of the wave.

10. The distance between five troughs of a wave is 4m. If the period of the wave is 0.8sec, find:

a. the wavelength of the wave

b. the velocity of the wave

c. the equation of the wave if its amplitude is 1.5m

CHAPTER 5
ECHOES

Echo is a sound heard after the reflection of sound wave from a plane surface. Echo can be used to determine the speed of sound in air by using the expression:

$$v = \frac{2x}{t},$$

where v is the speed of sound in air, x is the distance between the source of sound and the reflecting surface, while t is the time in seconds taken to hear the echo.

The speed of sound in air is also proportional to the square root of the air's absolute temperature. This shows that:

$$v \, \alpha \, \sqrt{T}$$

Examples

1. A man stands in front of a cliff and fires a gun. He hears the echo from the cliff after 2.5sec. If the speed of sound in air is 340m/s, how far is the man from the cliff?

   <u>Solution</u>

   $v = \frac{2x}{t}$ (Where x is the man's distance from the cliff)

   $340 = \frac{2x}{2.5}$

   $\therefore \ 2x = 340 \times 2.5$

   $2x = 850$

   $x = \frac{850}{2}$

   $x = 425m$

The man is at a distance of 425m from the cliff

2. A horn is sounded from a wall, 1320m way. How long will it take to hear the reflected sound? (Speed of sound in air is 330m)

Solution

$$v = \frac{2x}{t}$$

$$330 = \frac{2 \times 1320}{t}$$

$$330t = 2640$$

$$\therefore t = \frac{2640}{330}$$

$$t = 8.5 \text{sec}.$$

It will take 8.5sec to hear the reflected sound

3. A whistle is blown at a distance of 252m from a vertical wall. If the echo of the sound produced by the whistle is heard after 1.5sec, calculate the speed of sound in air.

Solution

$$v = \frac{2x}{t}$$

$$= \frac{2 \times 252}{1.5}$$

$$= \frac{504}{1.5}$$

$$\therefore v = 336 \text{ms}^{-1}$$

The speed of sound in air is 336ms^{-1}

4. Two people Jane and Michael are 170m apart along a horizontal ground. A vertical wall is 510m behind Michael. Jane fires a gun. What is the time interval between the two sounds:

a. heard by Michael

b. heard by Jane.

(Speed of sound in air is 340m)

Solutions

a. By the time Michael hears the reflected sound, the sound has travelled from Jane to Michael (170m), from Michael to the wall (510m) and from the wall to Michael (510m).

This gives a total distance of:

170 + 510 + 510 = 1190

$$\text{Velocity} = \frac{\text{Distance}}{\text{time}}$$

$$340 = \frac{1190}{t}$$

340t = 1190

∴ $t = \frac{1190}{340}$

t = 3.5sec

The time interval between the two sounds (first sound and echo) heard by Michael is 3.5sec.

b. The distance of Jane from the wall is 170 + 510 = 680m

$$v = \frac{2x}{t}$$ (Where x is Jane's distance from the wall)

$$340 = \frac{2 \times 680}{t}$$

340t = 1360

$$\therefore t = \frac{1360}{340} = 4$$

The time interval between the two sounds heard by Jane is 4sec.

5. A horn is sounded at regular intervals in front of a vertical wall 510m away. If the echo from the wall is heard simultaneously with the next hoot, how many hoots are made every minute. (Velocity of sound in air = 340ms^{-1})

Solution

$$v = \frac{2x}{t}$$

$$340 = \frac{2 \times 510}{t}$$

$$340t = 2 \times 510$$

$$t = \frac{1020}{340} = 3$$

The echo is heard every 3 seconds.

This is also the time taken to make each hoot. So, by simple proportion the number of hoots made in one minute is given by:

$$\frac{60}{3} = 20$$

The number of hoots made every minute is 20 hoots.

6. A boy 41m from a tall building claps his hands once every half second. He hears the echo of each clap midway between the clap and the next clap. Calculate the speed of sound.

Solution

The boy claps his hands every half second, i.e. $\frac{1}{2}$ second. He hears the echo midway between this time, i.e. $\frac{1}{2} \times (\frac{1}{2}$ second$) = \frac{1}{2} \times \frac{1}{2} = \frac{1}{4}$ seconds

So, it takes him $\frac{1}{4}$ seconds to hear the echo.

But, $v = \frac{2x}{t}$

$$= \frac{2 \times 41}{\frac{1}{4}} = \frac{82}{0.25} = 328$$

The speed of sound is 328ms^{-1}

7. The velocity of sound in air at a temperature of 25°C is 332ms^{-1}. What is the velocity when the temperature is 37°C?

Solution

In order to convert Celsius to Kelvin, add 273.

$v \propto \sqrt{T}$

$\therefore v = k\sqrt{T}$ (where k is a constant)

Substituting v = 332ms^{-1}, and T = 25 + 273 = 298k, gives:

$332 = k\sqrt{298}$

$\therefore k = \frac{332}{\sqrt{298}}$

When T = 37 + 273 = 310k, then v is obtained as follows:

$v = k\sqrt{T}$

$= \frac{332}{\sqrt{298}} \times \sqrt{310}$ ($k = \frac{332}{\sqrt{298}}$ from above)

$= 332\sqrt{\frac{310}{298}}$

$= 332 \times \sqrt{1.0403}$

= 332 x 1.02 = 338.6

The velocity at 37°C is 338.6ms^{-1}

8. The velocity of sound in air at a certain temperature is 344ms^{-1}. What is the velocity when the absolute temperature is increased by 16%?

Solution

If the initial temperature is T_1, then increasing T_1 by 16% gives,

$T_1 + \frac{16}{100}T_1 = T_1 + 0.16T_1 = 1.16T_1$

$v \propto \sqrt{T}$

∴ $v = k\sqrt{T}$ (where k is a constant)

Substituting v = 344ms^{-1}, and T = T_1, gives,

$344 = k\sqrt{T_1}$ ....Equation 1

Substituting v = v_2, and T = T_2 = 1.16T_1 gives:

$v_2 = k\sqrt{1.16T_1}$ ....Equation 2

Equation 2 divided by equation 1 gives:

$\frac{v_2}{344} = \frac{k\sqrt{1.16T_1}}{k\sqrt{T_1}}$

$\frac{v_2}{344} = \sqrt{\frac{1.16T_1}{T_1}}$ (k cancels out)

$\frac{v_2}{344} = \sqrt{1.16}$ (T_1 cancels out)

$\frac{v_2}{344} = 1.077$

∴ $v_2 = 344 \times 1.077 = 370.5$

The velocity when the temperature is increased by 16% is 370.5ms^{-1}

Exercise 5

1. A man stands in front of a mountain and fires a gun. He hears the echo from the mountain after 4sec. If the speed of sound in air is 330m/s, how far is the man from the mountain?

2. A horn is sounded from a tall building, 1200m way. How long will it take to hear the echo? (Speed of sound in air is 340m)

3. A whistle is blown at a distance of 332m from a vertical wall. If the echo of the sound produced by the whistle is heard after 2sec, calculate the speed of sound in air.

4. Two people P and Q are 220m apart along a horizontal ground. A cliff is 180m behind Q. P blows a whistle. What is the time interval between the two sounds:

a. heard by Q

b. heard by P?

(Speed of sound in air is 340m)

5. A horn is sounded at regular intervals in front of a vertical wall 200m away. If the reflected sound from the wall is heard simultaneously with the next hoot, how many hoots are made every minute. (Velocity of sound in air = 330ms^{-1})

6. A boy 162m from a cliff blows a whistle once every 2 seconds. He hears the echo of each whistle midway between the whistle and the next whistle. Calculate the speed of sound in air.

7. The velocity of sound in air at a temperature of 20°C is 328ms^{-1}. What is the velocity when the temperature is 39°C?

8. The velocity of sound in air at a certain absolute temperature is 338ms^{-1}. What is the velocity when the absolute temperature is increased by 25%?

CHAPTER 6
BEAT

Beat is a phenomenon which occurs when there is a rise and fall in the loudness of sound when two notes of nearly equal frequency are sounded together.

If f_1 and f_2 are the frequencies of the two notes with f_1 greater than f_2 then the beat frequency is given by:

$$f = f_1 - f_2$$

The beat frequency also means the number of beats made per second.

The period of the beat is given by:

$$T = \frac{1}{f}$$

Examples

1. Two notes of frequencies 30Hz and 25Hz are sounded together. Calculate:

    a. the beat frequency
    b. the period of the beat

    Solution

    a. $f = f_1 - f_2$ (Where f_1 is the larger frequency)

    $= 30 - 25 = 5Hz$

    $\therefore f = 5Hz$

    b. Recall that, $T = \frac{1}{f}$

    $\therefore T = \frac{1}{5} = 0.2sec$

    The period of the beat is 0.2sec

2. Two notes sounded together produced a beat of period 0.25sec. If the lower note has a frequency of 452Hz, calculate:

    a. the beat frequency
    b. the frequency of the higher note

Solutions

    a. The period is given by:

$$T = \frac{1}{f}$$

$$\therefore f = \frac{1}{T}$$

$$= \frac{1}{0.25} = 4$$

The beat frequency is 4Hz

    b. $f = f_1 - f_2$ Where f is the beat frequency and f_1 is the larger frequency of the two notes.

$$4 = f_1 - 452$$

$$\therefore 4 + 452 = f_1$$

$$456 = f_1$$

$$f_1 = 456$$

The frequency of the higher note is 456Hz

3. A note of 62Hz is sounded together with a note of 66Hz. How many beats per second will be heard?

Solution

The number of beats heard, $f = f_1 - f_2$

$$= 66 - 62 = 4Hz$$

The number of beats that will be heard is 4 beats per second.

Exercise 6

1. Two notes of frequencies 440Hz and 444Hz are sounded together. Calculate:
a. the beat frequency
b. the period of the beat

2. Two notes sounded together produced a beat of period 0.8sec. If the lower note has a frequency of 52Hz, calculate:
a. the beat frequency
b. the frequency of the higher note

3. A note of 100Hz is sounded together with a note of 106Hz. How many beats per second will be heard?

4. A beat per second of 3Hz was heard after sounding two notes. If the note of one of the sounds is 20Hz, calculate the two possible values of the other note.

5. Two notes sounded together produced a beat of period 1.4sec. If the higher note has a frequency of 325Hz, calculate:
a. the beat per second heard
b. the frequency of the lower note

CHAPTER 7
VIBRATION OF AIR COLUMN IN PIPES

1. Vibration in a closed pipe

At first resonance the length of air column in a closed pipe is given by:

$$l + c = \lambda/4$$

where l is the distance of the water level below the pipe, and c is the end correction which is a little distance of the wave above the top of the pipe. The value of c is usually negligible thereby giving the length of the air column as:

$$l = \lambda/4$$

Or, $\quad \lambda = 4l$

Therefore the resonant frequency of vibration is:

$$f_0 = v/\lambda \quad \text{(from } v = f\lambda\text{)}$$

$$= \frac{v}{4l}, \text{ where v is the speed of sound in air, } f_0 \text{ is the fundamental frequency of the closed pipe.}$$

The length of air column where the next resonance is observed in a closed pipe is given by:

$$l = 3\lambda/4$$

Or, $\quad \lambda = \dfrac{4l}{3}$

Therefore the frequency of vibration is given by:

$$f_1 = v/\lambda$$

$$= v \div \frac{4l}{3} \quad \text{(Since } \lambda = \frac{4l}{3}\text{)}$$

$$f_1 = \frac{3v}{4l}$$

$$f_1 = 3f_0 \quad \text{(Since } f_0 = \frac{v}{4l}\text{)}$$

This frequency, $f_1 = 3f_0$, is called the third harmonic or first overtone of a closed pipe. f_0 is the first harmonic. Other frequencies that can be obtained in a closed pipe include: $f_2 = 5f_0$, which is the fifth harmonic or second overtone, $f_3 = 7f_0$ and so on.

2. Vibration in an open pipe

At first resonance the length of air column in a open pipe is given by:

$$l = \lambda/2$$

Or, $\lambda = 2l$

Therefore the resonant frequency of vibration is:

$$f_0 = \frac{v}{2l},$$

The length of air column where the next resonance is observed (i.e. at the second harmonic or first overtone) in an open pipe is given by:

$$l = \lambda$$

Therefore the frequency of vibration is given by:

$$f_1 = \frac{v}{l}$$

$$= 2(\frac{v}{2l}), \quad \text{(Note that } 2(\frac{v}{2l}) = \frac{v}{l}\text{)}$$

$$f_1 = 2f_0 \quad \text{(Since } \frac{v}{2l} = f_0\text{)}$$

Similarly, the third harmonic, $f_2 = 3f_0$.

This shows that for an open pipe, $f_1 = 2f_0$, which is the second harmonic or first overtone, $f_2 = 3f_0$, which is the third harmonic or second overtone and so on. Therefore all harmonics are possible in an open pipe.

Note that: At first overtone, $l = \lambda$, at second overtone, $l = 3\lambda/2$, at third overtone, $l = 2\lambda$, and so on.

Examples

1. Calculate the frequency of fundamental of a closed pipe of length 20cm, if the speed of sound in air is 340m/s

   Solution

   $l = \lambda/4$ (The length of the first harmonic of a closed pipe, where λ is the wavelength)

   $\therefore \lambda = 4l$

   But, $l = 20cm = (\frac{20}{100})$ m $= 0.2$m

   $\lambda = 4 \times 0.2 = 0.8$m

   Recall that, $v = f\lambda$

   $\therefore f = v/\lambda$

   $= \frac{340}{0.8} = 425$

   $f = 425$Hz

2. If the length of the air column in a closed pipe is 45cm when it experiences its first overtone, the wavelength of the note is what? Calculate the frequency of the second overtone if the speed of sound in air is 340m/s

Solution

$l = 3\lambda/4$ (The length of the first overtone of a closed pipe)

$$\therefore \lambda = \frac{4l}{3}$$

But, $l = 45\text{cm} = \left(\frac{45}{100}\right)\text{m} = 0.45\text{m}$

$$\lambda = \frac{4 \times 0.45}{3}$$

$$= \frac{1.8}{3}$$

$$= 0.6\text{m}$$

The first overtone wavelength of the note is 0.6m

Recall that, $f = v/\lambda$ (Where $f = f_1$ = the frequency of the first overtone)

$$\therefore f_1 = v/\lambda$$

$$= \frac{340}{0.6} = 566.7$$

$f_1 = 566.7\text{Hz}$

But, $f_1 = 3f_0$

$$\therefore f_0 = \frac{f_1}{3}$$

$$= \frac{566.7}{3}$$

$f_0 = 188.9\text{Hz}$

But, $f_2 = 5f_0$ (Where f_2 is the frequency of the second overtone)

∴ $f_2 = 5 \times 188.9 = 944.5$

The frequency of the second overtone $f_2 = 944.5 Hz$

3. If the length of the air column in a closed pipe is 60cm when it experiences its third overtone, what is the wavelength of the note? Calculate the frequency of the first overtone if the speed of sound in air is 340m/s

Solution

$l = 7\lambda/4$ (The length of the third overtone of a closed pipe)

∴ $\lambda = \dfrac{4l}{7}$

But, $l = 60cm = \left(\dfrac{60}{100}\right) m = 0.60m$

$\lambda = \dfrac{4 \times 0.6}{7}$

$= \dfrac{2.4}{7}$

$= 0.343m$

The third overtone wavelength of the note is 0.343m

Recall that, $f = v/\lambda$ (Where $f = f_3 =$ the frequency of the third overtone)

∴ $f_3 = v/\lambda$

$= \dfrac{340}{0.343} = 991.3$

$f_3 = 991.3 Hz$

But, $f_3 = 7f_0$

$$\therefore f_o = \frac{f_3}{7}$$

$$= \frac{991.3}{7}$$

$f_o = 141.6 Hz$

But, $f_1 = 3f_o$ (Where f_1 is the frequency of the first overtone)

$\therefore f_1 = 3 \times 141.6 = 424.8$

The frequency of the first overtone $f_1 = 424.8 H_z$

4. The length of the air column in an open pipe is 108cm when it attains its first overtone. Calculate the frequency of the third overtone if the speed of sound in air is 330m/s

Solution

$l = \lambda$ (The length of the first overtone of an open pipe)

$\therefore \lambda = l$

But, $l = 108cm = (\frac{108}{100})$ m = 1.08m

$\lambda = 1.08m$

Recall that, $f = v/\lambda$ (Where $f = f_1$ = the frequency of the first overtone)

$\therefore f_1 = v/\lambda$

$= \frac{330}{1.08} = 305.6$

$f_1 = 305.6 Hz$

But, $f_1 = 2f_o$ (First overtone of an open tube)

$$\therefore f_0 = \frac{f_1}{2}$$

$$= \frac{305.6}{2}$$

$$f_0 = 152.8 Hz$$

But, $f_3 = 4f_0$ (Where f_3 is the frequency of the third overtone)

$\therefore \quad f_3 = 4 \times 152.8 = 611.2$

The frequency of the third overtone $f_3 = 611.2 Hz$

5. What is the shortest length of an opened tube which resonates with a frequency of 320Hz? (speed of sound in air = 330m/s)

Solution

$l = \lambda/2$ (Shortest length is the length of the first harmonics)

$\therefore \lambda = 2l$

But, $v = f\lambda$

$v = f \times 2l$ (Since $\lambda = 2l$)

$\therefore 2fl = v$

$$l = \frac{v}{2f}$$

$$l = \frac{330}{2 \times 320}$$

$$= \frac{330}{640} = 0.516m$$

6. The length of the vibrating air column of a simple resonance tube can be altered by adjusting a water level. Resonance is found for a tuning fork of frequency 440Hz when the length of the air column is 18.8cm and again when it is 57.3cm. Calculate the speed of sound of the air in the tube.

Solution

Assuming the two positions to be the first and second resonance positions, then:

$l_1 = \lambda/4$ (Position of first resonance)

$l_2 = 3\lambda/4$ (Position of second resonance)

$l_2 - l_1 = 3\lambda/4 - \lambda/4$

$l_2 - l_1 = 2\lambda/4$

$l_2 - l_1 = \lambda/2$ (When expressed in its lowest term)

$57.3 - 18.8 = \lambda/2$

$38.5 = \lambda/2$

$\therefore \lambda = 2 \times 38.5 = 77$cm

$\lambda = (\frac{77}{100})m = 0.77m$

But, $v = f\lambda$

$v = 440 \times 0.77$ (f = 440, as given in the question)

$v = 338.8 ms^{-1}$

The speed of sound of the air in the tube is $338.8 ms^{-1}$

7. In a resonance tube experiment, resonance is found for a tuning fork of frequency 480Hz when the length of the air column is 24.7cm and again when the air column is 59.9cm long. Calculate the speed of sound in air.

Solution

As shown in example (6) above, the difference in length of air column between any two consecutive positions of resonance in a closed pipe is $\lambda/2$.

(e.g. $5\lambda/4 - 3\lambda/4 = \lambda/2$)

$\therefore 59.9 - 24.7 = \lambda/2$

$35.2 = \lambda/2$

$\therefore \lambda = 2 \times 35.2 = 70.4\text{cm}$

$\lambda = (\frac{70.4}{100})\text{m} = 0.704\text{m}$

But, $v = f\lambda$

$\therefore v = 480 \times 0.704$ (f = 480, as given in the question)

$v = 337.9\text{ms}^{-1}$

The speed of sound in air is 337.9ms^{-1}

Exercise 7

1. Calculate the frequency of fundamental of a closed pipe of length 42cm, if the speed of sound in air is 330m/s

2. If the length of the air column in an open pipe is 45cm when it experiences its first overtone, what is the wavelength of the note? Calculate the frequency of the third overtone if the speed of sound in air is 340m/s

3. If the length of the air column in a closed pipe is 25cm when it experiences its second overtone, what is the wavelength of the note? Calculate the frequency of the first overtone if the speed of sound in air is 340m/s

4. The length of the air column in an open pipe is 75cm when it attains its second overtone. Calculate the frequency of the fifth overtone if the speed of sound in air is 330m/s

5. What is the shortest length of an opened tube which resonates with a frequency of 82Hz? (Speed of sound in air = 330m/s)

6. The length of the vibrating air column of a simple resonance tube can be altered by adjusting a water level. Resonance is found for a tuning fork of frequency 380Hz when the length of the air column is 24.2cm and again when it is 72.1cm. Calculate the speed of sound of the air in the tube.

7. In a resonance tube experiment, resonance is found for a tuning fork of frequency 560Hz when the length of the air column is 16.5cm and again when the air column is 46.3cm long. Calculate the speed of sound in air.

8. The length of the air column in an open pipe is 0.96m when it attains its third overtone. Calculate the frequency of the first overtone if the speed of sound in air is 340m/s

CHAPTER 8
MODES OF VIBRATION OF A STRETCHED STRING

The simplest mode of vibration (fundamental) of a plucked stretched string is given by:

$$l = \lambda/2$$

$$\therefore \lambda = 2l$$

The fundamental frequency is thus given by:

$$f_0 = \frac{v}{2l} \quad \text{(from } v = f\lambda\text{)}$$

At first overtone, $l = \lambda$, at second overtone, $l = 3\lambda/2$, at third overtone, $l = 2\lambda$, and so on.

As with an open pipe, all possible harmonics, f_0, $f_1=2f_0$, $f_2=3f_0$, $f_3=4f_0$, etc, can be obtained on a stretched string.

The velocity of a wave propagated along a fixed wire or string is given by:

$$v = \sqrt{\frac{T}{m}}$$

The fundamental frequency can therefore be given by:

$$f_0 = \frac{1}{2l}\sqrt{\frac{T}{m}} \quad \text{(from } f = v/\lambda\text{)}, \text{ where T is the tension in the string,}$$

m is the mass per unit length of the string in kg/m, and l is the length of the string in metre. This shows that:

$f \alpha \sqrt{T}$ (when l and m are constant)

$f \alpha \frac{1}{l}$ (when T and m are constant)

$f \alpha \dfrac{1}{\sqrt{m}}$ (when T and *l* are constant)

Examples

1. The frequency of a standard string when the tension on the string is 20N was found to be 45Hz. Determine the frequency when the tension is increased to 80N.

Solution

$\qquad f \alpha \sqrt{T}$ (f is the frequency while T is the tension)

$\therefore f = k\sqrt{T}$ (Where k is a constant)

When f = 45 and T = 20, then:

$45 = k\sqrt{20}$

$k = \dfrac{45}{\sqrt{20}}$

When T = 80, then f is obtained as follows:

$f = k\sqrt{T}$

$= \dfrac{45}{\sqrt{20}} \times \sqrt{80}$ (Since $k = \dfrac{45}{\sqrt{20}}$)

$= 45\sqrt{\dfrac{80}{20}}$

$= 45\sqrt{4}$

$= 45 \times 2 = 90$

The frequency is 90Hz

2. The frequency of a guitar string when the tension on the string is 42N was found to be 60Hz. Determine the frequency of the string when the tension is decreased to 12N.

Solution

$$f \alpha \sqrt{T}$$

$$\therefore f = k\sqrt{T} \quad \text{(Where k is a constant)}$$

When f = 60 and T = 42, then:

$$60 = k\sqrt{42}$$

$$k = \frac{60}{\sqrt{42}}$$

When T = 12, then f is obtained as follows:

$$f = k\sqrt{T}$$

$$= \frac{60}{\sqrt{42}} \times \sqrt{12} \quad \text{(Since k = }\frac{60}{\sqrt{42}}\text{)}$$

$$= 60\sqrt{\frac{12}{42}}$$

$$= 60\sqrt{0.2857}$$

$$= 60 \times 0.5345$$

$$= 32.1 \text{Hz}$$

3. A string has a frequency of 54Hz when a tension of 28N is applied on it. Determine the new tension on the string when the frequency is 20Hz.

Solution

Another method of solving a question like this is as explained below.

$$f = k\sqrt{T} \quad \text{(Where k is a constant)}$$

When f = 54 and T = 28, then:

$54 = k\sqrt{28}$ .... Equation (1)

When f = 20 and T = T_2, then:

$20 = k\sqrt{T_2}$ .... Equation (2) (T_2 is the unknown tension to be calculated)

Divide equation (1) by equation (2). This gives:

$$\frac{54}{20} = \frac{k\sqrt{28}}{k\sqrt{T_2}}$$

(k cancels out to give)

$$\therefore \frac{54}{20} = \sqrt{\frac{28}{T_2}}$$

$$2.7 = \sqrt{\frac{28}{T_2}}$$

Square both sides. This gives:

$2.7^2 = \frac{28}{T_2}$ (When $\sqrt{\frac{28}{T_2}}$ is squared, the square root sign clears out)

$\therefore 7.29 = \frac{28}{T_2}$

$7.29 T_2 = 28$

$T_2 = \frac{28}{7.29} = 3.84$

The new tension on the string is 3.84N

4. The frequency of a string emitting a note is 50Hz. Determine its frequency when the tension is halved.

<u>Solution</u>

Let the original and new tensions be T_1 and T_2 respectively. (But $T_2 = \dfrac{T_1}{2}$ since the new tension is half of the original tension)

$f = k\sqrt{T}$ (Where k is a constant)

When $f = 50$ and $T = T_1$, then:

$50 = k\sqrt{T_1}$ .... Equation (1)

When $f = f_2$ and $T = T_2 = \dfrac{T_1}{2}$, then:

$f_2 = k\sqrt{\dfrac{T_1}{2}}$ .... Equation (2)   (f_2 is the unknown frequency to be calculated)

Divide equation (1) by equation (2). This gives:

$$\dfrac{50}{f_2} = \dfrac{k\sqrt{T_1}}{k\sqrt{\dfrac{T_1}{2}}}$$

$\therefore \dfrac{50}{f_2} = \sqrt{\dfrac{T_1}{\dfrac{T_1}{2}}}$ (k cancels out)

$\dfrac{50}{f_2} = \sqrt{\dfrac{2T_1}{T_1}}$

$\dfrac{50}{f_2} = \sqrt{2}$ (T_1 cancels out)

$\dfrac{50}{f_2} = 1.414$

$\therefore 1.414 f_2 = 50$

$f_2 = \dfrac{50}{1.414} = 35.4$

The frequency is 35.4N

5. A string emits a note of frequency 100Hz when a certain tension was applied on it. Determine the frequency of the string when the tension is quadrupled.

Solution

Let the original and new tensions be T_1 and T_2 respectively. (But $T_2 = 4T_1$ since the new tension is four times the original tension i.e. quadrupled.

$f = k\sqrt{T}$

When f = 100 and T = T_1, then:

$100 = k\sqrt{T_1}$

∴ $k = \dfrac{100}{\sqrt{T_1}}$

When f = f_2 and T = $T_2 = 4T_1$, then:

$f_2 = k\sqrt{T_2}$

$f_2 = \dfrac{100}{\sqrt{T_1}} \times \sqrt{4T_1}$ (Since $k = \dfrac{100}{\sqrt{T_1}}$ and $T_2 = 4T_1$)

$f_2 = 100\sqrt{\dfrac{4T_1}{T_1}}$

T_1 cancels out to give:

$f_2 = 100\sqrt{4}$

$f_2 = 100 \times 2 = 200$

The frequency of the string is 200Hz

6. A string of length 42cm and mass 12g emits a note when a tension of 75N was applied on it. Determine the fundamental frequency of the string.

Solution

Given: T = 75N, l = 42cm = $(\frac{42}{100})$m = 0.42m

m = mass per unit length. But mass of wire = $(\frac{12}{1000})$kg = 0.012kg

∴ mass per unit length, m = $\frac{0.012}{0.42}$ = 0.02857kg/m

So, the fundamental frequency is given by:

$$f_0 = \frac{1}{2l}\sqrt{\frac{T}{m}}$$

$$= \frac{1}{2 \times 0.42}\sqrt{\frac{75}{0.02857}}$$

$$= \frac{1}{0.84}\sqrt{2625}$$

$$= \frac{1}{0.84} \times 51.23 = \frac{51.23}{0.84} = 60.988$$

The fundamental frequency, f_0 is 61.0Hz

7. A string of length 80cm and mass 4g emits a note when a tension of 20N was applied on it. Determine:

a. the frequency of the third harmonic

b. the frequency of the fifth overtone

c. the velocity of the wave in the string

Solution

a. An easy method of solving this question is to first determine the fundamental frequency, f_0.

Given: T = 20N, l = 80cm = $(\frac{80}{100})$m = 0.8m

m = mass per unit length. But mass of wire = $(\frac{4}{1000})$kg = 0.004kg

∴ mass per unit length, $m = \dfrac{0.004}{0.8} = 0.005 \text{ kg/m}$

So, the fundamental frequency is given by:

$$f_0 = \dfrac{1}{2l}\sqrt{\dfrac{T}{m}}$$

$$= \dfrac{1}{2 \times 0.8}\sqrt{\dfrac{20}{0.005}}$$

$$= \dfrac{1}{1.6}\sqrt{4000}$$

$$= \dfrac{1}{1.6} \times 63.25$$

$$= \dfrac{63.25}{1.6} = 39.5$$

The fundamental frequency, f_0 is 39.5Hz

∴ The frequency of the third harmonic = $3f_0$ = 3 × 39.5 = 118.5 (Note that harmonics are multiples of the fundamental frequency)

Hence, the frequency of the third harmonic is 118.5Hz

b. The frequency of the fifth overtone, $f_5 = 6f_0$

∴ f_5 = 6 × 39.5 = 237

The frequency of the fifth overtone is 237Hz.

c. The velocity of the wave is given by:

$$v = \sqrt{\dfrac{T}{m}}$$

$$= \sqrt{\dfrac{20}{0.005}}$$

$$= \sqrt{4000}$$

$$= 63.25$$

∴ The velocity of the wave is 63.25ms^{-1}

8. A string when plucked emits a note of frequency 36Hz. Determine the frequency of the string when the length is tripled, and the tension on the string is kept constant.

Solution

Let the original and new lengths be l_1 and l_2 respectively. (But $l_2 = 3l_1$ since the new length is three times the original length)

∴ $f \alpha \dfrac{1}{l}$ (when T and m are constant)

$f = \dfrac{k}{l}$ (where k is a constant)

When $f = f_1 = 36\text{Hz}$, and $l = l_1$ then f_1 is given by:

$$f_1 = \dfrac{k}{l_1}$$

$$36 = \dfrac{k}{l_1}$$

∴ $k = 36l_1$

When $f = f_2$, and $l = l_2 = 3l_1$ then f_2 is given by:

$$f_2 = \dfrac{k}{l_2}$$

$$= \dfrac{36l_1}{3l_1} \quad \text{(Since } k = 36l_1\text{)}$$

$$= 12 \quad (l_1 \text{ cancels out})$$

The frequency is 12Hz when the length is tripled.

9. When a musical string is plucked, a note of frequency 120Hz is emitted. Determine the frequency of the string when its mass per unit length is halved, while the tension on the string and length of the string are kept constant.

Solution

Let the original and new mass per unit lengths be m_1 and m_2 respectively. (But $m_2 = \frac{m_1}{2}$ since the new mass per unit length is half times the original value)

$$\therefore f \alpha \frac{1}{\sqrt{m}} \quad \text{(when T and } l \text{ are constant)}$$

$$f = \frac{k}{\sqrt{m}} \quad \text{(where k is a constant)}$$

When $f = f_1 = 120$Hz, and $m = m_1$ then f_1 is given by:

$$f_1 = \frac{k}{\sqrt{m_1}}$$

$$120 = \frac{k}{\sqrt{m_1}}$$

$$\therefore k = 120\sqrt{m_1}$$

When $f = f_2$, and $m = m_2 = \frac{m_1}{2}$ then f_2 is given by:

$$f_2 = \frac{k}{\sqrt{m_1}}$$

$$= \frac{120\sqrt{m_1}}{\sqrt{\frac{m_1}{2}}} \quad \text{(Since k = } 120\sqrt{m_1}\text{)}$$

$$= 120\sqrt{m_1} \times \sqrt{\frac{2}{m_1}}$$

$$= 120 \times \sqrt{\frac{2m_1}{m_1}}$$

$$= 120 \times \sqrt{2} \quad (m_1 \text{ cancels out})$$

$$= 120 \times 1.414$$

$$= 169.7$$

The frequency is 169.7Hz.

10. A plucked string of length 70cm emits a note of certain frequency. Determine the new length of the string when the frequency is increased to five times its original value, and the tension on the string is kept constant.

Solution

Let the original and new frequencies be f_1 and f_2 respectively. (But $f_2 = 5f_1$ since the new frequency is five times the original frequency)

$$\therefore f \alpha \frac{1}{l} \quad \text{(when T and m are constant)}$$

$$f = \frac{k}{l} \quad \text{(where k is a constant)}$$

When $f = f_1$, and $l = l_1 = 70$cm then f_1 is given by:

$$f_1 = \frac{k}{l_1}$$

$$f_1 = \frac{k}{70}$$

$$\therefore k = 70f_1$$

When $f = f_2 = 5f_1$, and $l = l_2$, then f_2 is given by:

$$f_2 = \frac{k}{l_2}$$

$5f_1 = \dfrac{70f_1}{l_2}$ (Since k = 70f₁ and f₂ = 5f₁)

$5f_1 l_2 = 70f_1$

$\therefore l_2 = \dfrac{70f_1}{5f_1}$

= 14   (f₁ cancels out)

The new length is 14Hz when the frequency is increased five times.

Exercise 8

1. The frequency of a standard string when the tension on the string is 32N was found to be 50Hz. Determine the frequency when the tension is increased to 60N.

2. The frequency of a violin string when the tension on the string is 100N was found to be 85Hz. Determine the frequency of the string when the tension is increased to 140N.

3. A string has a frequency of 300Hz when a tension of 120N is applied on it. Determine the new tension on the string when the frequency becomes 80Hz.

4. The frequency of a string emitting a note is 50Hz. Determine its frequency when the tension becomes five times its original value.

5. A string emits a note of frequency 420Hz when a certain tension was applied on it. Determine the frequency of the string when the tension is halved.

6. A string of length 0.65m and mass 10g emits a note when a tension of 100N was applied on it. Determine the fundamental frequency of the string.

7. A string of length 210cm and mass 20g emits a note when a tension of 32N was applied on it. Determine:

a. the frequency of the second harmonic

b. the frequency of the fourth overtone

c. the velocity of the wave in the string

8. A string when plucked emits a note of frequency 220Hz. Determine the frequency of the string when the length is quadrupled, and the tension on the string remains the same.

9. When a musical string is plucked, a note of frequency 25Hz is emitted. Determine the frequency of the string when its mass per unit length is tripled, while the tension on the string and length of the string are kept constant.

10. A plucked string of length 100cm emits a note of certain frequency. Determine the new length of the string when the frequency is doubled and the tension on the string is kept constant.

CHAPTER 9
CHARACTERISTICS OF SOUND – THE PITCH

The pitch of a sound depends on the frequency of the sound wave. Two methods which can be used to show that the pitch of a sound depends on its frequency are:

1. The use of disc siren. This method shows that the frequency of the wave produced is given by:

$$f = \frac{\text{no. of revolutions } \times \text{ no. of holes on the disc}}{\text{time of revolution in seconds}}$$

2. The use of toothed wheel. This method shows that the frequency of the wave produced is given by:

$$f = \frac{\text{no. of revolutions } \times \text{ no. of teeth on the wheel}}{\text{time of revolution in seconds}}$$

Examples

1. An experiment was carried out to determine the frequency of a note. A disc siren was observed to make a revolution of 360 for a period of 3 minutes. If the disc is made up of 40 evenly spaced holes, what is the frequency of the note emitted?

Solution

$$f = \frac{\text{no. of revolutions } \times \text{ no. of holes on the disc}}{\text{time of revolution in seconds}}$$

$$= \frac{360 \times 40}{3 \times 60}$$ (The 3 minutes is multiplied by 60 to convert it to seconds)

$= 20 \times 4$ (After equal divisions by 10, 6 and then 3)

∴ f = 80 vibrations per second or 80Hz

2. A disc siren produced a note of frequency 240Hz. If the disc is made up of 60 evenly spaced holes, what is the speed of the disc?

Solution

Let the speed be the number of revolutions made per second, i.e. in 1 second. So, the time is 1 second.

$$f = \frac{\text{no. of revolutions} \times \text{no. of holes on the disc}}{\text{time of revolution in seconds}}$$

$$240 = \frac{\text{no. of revolutions} \times 60}{1}$$

$$240 = \frac{\text{no. of revolutions}}{1} \times 60$$

∴ No. of revolutions/sec = $\frac{240}{60}$

= 4

(Note that $\frac{\text{no. of revolutions}}{1}$ is the no of revolutions/sec)

∴ Speed of the disc is 4revs./sec

3. In an experiment to produce a note of frequency 300Hz, a toothed wheel was rotated to make 250 revolutions in 2 minutes. Calculate the number of teeth on the wheel.

Solution

$$f = \frac{\text{no. of revolutions} \times \text{no. of teeth on the wheel}}{\text{time of revolution in seconds}}$$

$$300 = \frac{250 \times \text{no. of teeth}}{2 \times 60}$$ (The 2 minutes is multiplied by 60 to convert it to seconds)

300 x 2 x 60 = 250 x no. of teeth

∴ No. of teeth = $\dfrac{300 \times 2 \times 60}{250}$

= 12 x 2 x 6   (After equal divisions by 10 and 25)

= 144

The wheel has 144 teeth.

4. A note of frequency 200Hz was produced by a toothed wheel. If the wheel has 60 teeth, calculate the time taken by the wheel to complete 240 revolutions.

Solution

$$f = \dfrac{\text{no. of revolutions } \times \text{ no. of teeth on the wheel}}{\text{time of revolution in seconds}}$$

$$200 = \dfrac{240 \times 60}{\text{time in seconds}}$$

200 x time = 240 x 60

∴ Time = $\dfrac{240 \times 60}{200}$

= 12 x 6   (After cancelling out the zeros and dividing by 2)

= 72 seconds

Time in minutes = $\dfrac{72}{60}$ = 1.2 minutes

It takes 1.2 minutes for the wheel to make the revolutions.

Exercise 9

1. An experiment was carried out to determine the frequency of a note. A disc siren was observed to make a revolution of 540 for a period of 4 minutes. If the disc is made up of 60 evenly spaced holes, what is the frequency of the note emitted?

2. A disc siren produced a note of frequency 44Hz. If the disc is made up of 50 evenly spaced holes, what is the speed of the disc?

3. In an experiment to produce a note of frequency 80Hz, a toothed wheel was rotated to make 320 revolutions in 2.5 minutes. Calculate the number of teeth on the wheel.

4. A note of frequency 540Hz was produced by a toothed wheel. If the wheel has 80 teeth, calculate the time taken by the wheel to complete 150 revolutions.

5. A note of frequency 100Hz was produced by a disc siren. If the disc has 40 evenly spaced holes, calculate the time taken by the disc to complete 380 revolutions.

CHAPTER 10
DOPPLER EFFECTS IN SOUND

Doppler effect is the change in frequency (pitch) of a source of sound when there is a relative motion between the source and an observer.

Expressions for apparent frequencies

Case 1   When the source of sound is moving towards a stationary observer, the apparent frequency is given by:

$$f' = \left(\frac{V}{V-V_s}\right)f$$

Where f' is the apparent frequency, V is the velocity of sound in air, V_s is the velocity of the source of sound, and f is the frequency emitted from the source of sound.

Case 2   When the source of sound is moving away from a stationary observer, the apparent frequency is given by:

$$f' = \left(\frac{V}{V+V_s}\right)f$$

Case 3   When the source of sound is stationary and an observer is moving towards it, then the apparent frequency is given by:

$$f' = \left(\frac{V+V_o}{V}\right)f \quad \text{where } V_o \text{ is the velocity of the observer.}$$

Case 4   When the source of sound is stationary and an observer is moving away from it, then the apparent frequency is given by:

$$f' = \left(\frac{V-V_o}{V}\right)f$$

Case 5   When both the source of sound and an observer are moving towards each other, then the apparent frequency is given by:

$$f' = \left(\frac{V+V_o}{V-V_s}\right)f$$

Case 6 When both the source of sound and an observer are moving away from each other, then the apparent velocity is given by:

$$f' = \left(\frac{V-V_o}{V+V_s}\right)f$$

Case 7 When the source of sound and an observer are moving in the same direction, then we can have two cases as stated below.

a. When the source is behind the observer, then the apparent frequency is given by:

$$f' = \left(\frac{V-V_o}{V-V_s}\right)f$$

b. When the source is in front of the observer, then the apparent frequency is given by:

$$f' = \left(\frac{V+V_o}{V+V_s}\right)f$$

Note that the quantities, $V-V_o$, $V+V_o$, $V-V_s$ and $V+V_s$ are the relative velocities between the velocity of sound in air and the velocity of the source of sound or the observer. When both velocities are moving in the same direction we subtract their velocities. When they are moving in opposite direction we add their velocities. All sounds are moving away from the source to the observer.

Examples

1. A train approaching a station with a speed of $25ms^{-1}$ sounds its horn and emits a note of frequency 450Hz. What frequency will be received at the station? (Speed of sound in air is $330ms^{-1}$).

Solution

This is case 1 as stated above. The apparent frequency is given by:

$$f' = \left(\frac{V}{V-V_s}\right)f$$

where V=330ms^{-1}, V$_s$ = 25ms^{-1} and f = 450Hz. Substituting these values into the expression above gives:

$$f' = (\frac{330}{330-25}) \times 450$$

$$= (\frac{330}{305}) \times 450$$

$$= 486.9$$

The station will receive a frequency of 486.9Hz.

2. A sounding tuning fork of frequency 420Hz is moved away from an observer with a speed of 4ms^{-1}. What is the apparent frequency of the sound coming to the observer? (Speed of sound in air is 330ms^{-1}).

Solution

This is case 2 above. The apparent frequency is given by:

$$f' = (\frac{V}{V+V_s})f$$

where V=330ms^{-1}, V$_s$ = 4ms^{-1} and f = 420Hz. Substituting these values into the expression above gives:

$$f' = (\frac{330}{330+4}) \times 420$$

$$= (\frac{330}{334}) \times 420$$

$$= 415$$

The frequency coming to the observer is 415Hz.

3. A stationary siren emits a note of frequency 382Hz. What frequency will be received by a train which is approaching it at a speed of 20ms^{-1}? (Speed of sound in air is 330ms^{-1}).

Solution

This is case 3 above. The apparent frequency is given by:

$$f' = \left(\frac{V+V_o}{V}\right)f$$

where $V = 330 ms^{-1}$, $V_o = 20 ms^{-1}$ and $f = 382 Hz$. Substituting these values into the expression above gives:

$$f' = \left(\frac{330+20}{330}\right) \times 382$$

$$= \left(\frac{350}{330}\right) \times 382$$

$$= 405.2$$

The train will receive a frequency of 405.2Hz.

4. A stationary siren at a train station emits a note of frequency 365Hz. What frequency will be received by a train which is leaving the station at a speed of $11 ms^{-1}$? (Speed of sound in air is $333 ms^{-1}$).

Solution

This is case 4 as stated above. The apparent frequency is given by:

$$f' = \left(\frac{V-V_o}{V}\right)f$$

where $V = 333 ms^{-1}$, $V_o = 11 ms^{-1}$ and $f = 365 Hz$. Substituting these values into the expression above gives:

$$f' = \left(\frac{333-11}{333}\right) \times 365$$

$$= \left(\frac{322}{333}\right) \times 365$$

$$= 352.9$$

The train will receive a frequency of 352.9Hz.

5. A student sounding a tuning fork of frequency 500Hz moves towards a man with a speed of 2ms^{-1}. If the man is running towards the student with a speed of 3ms^{-1}, what is the apparent frequency of the sound coming to the man? (Speed of sound in air is 333ms^{-1}).

Solution

This is case 5 above. The apparent frequency is given by:

$$f' = \left(\frac{V+Vo}{V-Vs}\right)f$$

where V=333ms^{-1}, V_s = 2ms^{-1}, V_o = 3ms^{-1} and f = 500Hz. Substituting these values into the expression above gives:

$$f' = \left(\frac{333+3}{333-2}\right) \times 500$$

$$= \left(\frac{336}{331}\right) \times 500$$

$$= 507.6$$

The frequency coming to the man is 507.6Hz.

6. A football referee and a footballer run away from each other on a football field. The referee runs with a velocity of 2ms^{-1} while the footballer runs with a velocity of 4ms^{-1}. If the referee blows his whistle and emits a note of frequency 340Hz, what is the apparent frequency of the sound that the footballer will hear? (Speed of sound in air is 331ms^{-1}).

Solution

This is case 6 above. The apparent frequency is given by:

$$f' = \left(\frac{V-Vo}{V+Vs}\right)f$$

where V=331ms^{-1}, V$_s$ = 2ms^{-1}, V$_o$ = 4ms^{-1} and f = 340Hz. Substituting these values into the expression above gives:

$$f' = (\frac{331-4}{331+2}) \times 340$$

$$= (\frac{327}{333}) \times 340$$

$$= 333.9$$

The frequency that the footballer will hear is 333.9Hz.

7. A football referee and a footballer are running in the same direction with the referee behind the footballer. The referee runs with a velocity of 3ms^{-1} while the footballer runs with a velocity of 2ms^{-1}. If the referee blows his whistle and emits a note of frequency 384Hz, what is the frequency of the sound coming to the footballer? (Speed of sound in air is 331ms^{-1}).

Solution

This is case 7a above. The apparent frequency is given by:

$$f' = (\frac{V-Vo}{V-Vs})f$$

where V=331ms^{-1}, V$_s$ = 3ms^{-1}, V$_o$ = 2ms^{-1} and f = 384Hz. Substituting these values into the expression above gives:

$$f' = (\frac{331-2}{331-3}) \times 384$$

$$= (\frac{329}{328}) \times 384$$

$$= 385.2$$

The frequency that the footballer will hear is 385.2Hz.

8. A car travelling at a speed of 30ms^{-1} passes a bicycle which is moving at a speed of 7ms^{-1} in the opposite direction. If the car immediately sounds its horn and emits a

note of frequency 480Hz, calculate the apparent frequency that will come to the bicycle? (Speed of sound in air is 331ms^{-1}).

Solution

This is case 7b above. The apparent frequency is given by:

$$f' = \left(\frac{V+V_o}{V+V_s}\right)f$$

where V=331ms^{-1}, V_s = 30ms^{-1}, V_o = 7ms^{-1} and f = 480Hz. Substituting these values into the expression above gives:

$$f' = \left(\frac{331+7}{331+30}\right) \times 480$$

$$= \left(\frac{338}{361}\right) \times 480$$

$$= 449.4$$

The frequency that will come to the bicycle is 449.4Hz.

9. A stationary siren emits a note of frequency 430Hz. If the apparent frequency received by a train which is approaching it at a speed of 12.5ms^{-1} is 446Hz, determine the speed of sound in air.

Solution

This is case 3 above. The apparent frequency is given by:

$$f' = \left(\frac{V+V_o}{V}\right)f$$

where f' = 446Hz, V_o = 12.5ms^{-1} and f = 430Hz. Substituting these values into the expression above gives:

$$446 = \left(\frac{V+12.5}{V}\right) \times 430$$

Dividing both sides by 430 gives:

$$\frac{446}{430} = \left(\frac{V+12.5}{V}\right)$$

$$1.037 = \frac{V+12.5}{V}$$

$$1.037V = V + 12.5$$

$$1.037V - V = 12.5$$

$$0.037V = 12.5$$

$$\therefore V = \frac{12.5}{0.037} = 337.8$$

The speed of sound in air is 337.8ms^{-1}.

10. A student moving towards his classmate blows a whistle and emits a note of frequency 250Hz. If his classmate running towards him at a speed of 5ms^{-1}, hears the sound at an apparent frequency of 256Hz, determine the velocity with which the student was moving towards his classmate. (Speed of sound in air is 330ms^{-1}).

Solution

This is case 5 above. The apparent frequency is given by:

$$f' = \left(\frac{V+Vo}{V-Vs}\right)f$$

where f'= 256Hz, V=330ms^{-1}, V$_o$ = 5ms^{-1} and f = 250Hz. Substituting these values into the expression above gives:

$$256 = \left(\frac{330+5}{330-Vs}\right) \times 250$$

Dividing both sides by 250 gives:

$$\frac{256}{250} = \left(\frac{335}{330-Vs}\right)$$

$$1.024 = \frac{335}{330-Vs}$$

$1.024(330 - V_s) = 335$

$337.92 - 1.024V_s = 335$

$337.92 - 335 = 1.024V_s$

$2.92 = 1.024V_s$

$\therefore V_s = \dfrac{2.92}{1.024} = 2.85$

The student was moving towards his classmate with a velocity of 2.85ms^{-1}

Exercise 10

1. A train approaching a station with a speed of 18ms^{-1} sounds its horn and emits a note of frequency 220Hz. What frequency will be received at the station? (Speed of sound in air is 340ms^{-1}).

2. A blown whistle of frequency 360Hz is moved away from an observer with a speed of 2.5ms^{-1}. What is the apparent frequency of the sound coming to an observer? (Speed of sound in air is 330ms^{-1}).

3. A stationary siren emits a note of frequency 430Hz. What frequency will be received by a train which is approaching it at a speed of 16ms^{-1}? (Speed of sound in air is 340ms^{-1}).

4. A stationary siren at a train station emits a note of frequency 384Hz. What frequency will be received by a train which is leaving the station at a speed of 8ms^{-1}? (Speed of sound in air is 340ms^{-1}).

5. A student sounding a tuning fork of frequency 450Hz moves towards a man with a speed of 3ms^{-1}. If the man is running towards the student with a speed of 4.5ms^{-1}, what is the apparent frequency of the sound coming to the man? (Speed of sound in air is 340ms^{-1}).

6. A football referee and a footballer run away from each other on a football field. The referee runs with a velocity of 1ms^{-1} while the footballer runs with a velocity of 2ms^{-1}. If the referee blows his whistle and emits a note of frequency 350Hz, what is the apparent frequency of the sound that the footballer will hear? (Speed of sound in air is 340ms^{-1}).

7. A football referee and a footballer are running in the same direction with the referee behind the footballer. The referee runs with a velocity of 4ms^{-1} while the footballer runs with a velocity of 2ms^{-1}. If the referee blows his whistle and emits a note of frequency 396Hz, what is the frequency of the sound coming to the footballer? (Speed of sound in air is 340ms^{-1}).

8. A car travelling at a speed of 30ms^{-1} passes a bike which is moving at a speed of 12ms^{-1} in the opposite direction. If the car immediately sounds its horn and emits a note of frequency 464Hz, calculate the apparent frequency that will come to the bike? (Speed of sound in air is 340ms^{-1}).

9. A stationary siren emits a note of frequency 421Hz. If the apparent frequency received by a train which is approaching it at a speed of 10ms^{-1} is 435Hz, determine the speed of sound in air.

10. A student moving towards his classmate blows a whistle and emits a note of frequency 220Hz. If his classmate running towards him at a speed of 4ms^{-1}, hears the sound at an apparent frequency of 225Hz, determine the velocity with which the student was moving towards his classmate. (Speed of sound in air is 340ms^{-1}).

CHAPTER 11
LINEAR EXPANSIVITY

When solids are heated, they increase in length. Also, when they are cooled, they decrease in length. This change in length can be expressed in terms of a quantity called linear expansivity. Linear expansivity can be expressed as follows:

$$\alpha = \frac{\Delta L}{L_1 \Delta \theta}$$

Or, $$\alpha = \frac{L_2 - L_1}{L_1(\theta_2 - \theta_1)}$$

Where α = linear expansivity, $\Delta L = (L_2 - L_1)$ = change in length, L_1 = the initial length, L_2 = final length. L_2 is greater than L_1. Also, $\Delta \theta = (\theta_2 - \theta_1)$ = change in temperature, and θ_2 is a higher temperature than θ_1.

The unit of linear expansivity is per Kelvin (K^{-1}) or /°C.

Examples

1. An iron rod has a length of 60m when the temperature is 15°C. By how much will it expand when the temperature rises to 40°C? (Linear expansivity of iron = 0.000012 K^{-1})

Solution

Given in the question are: L_1 = 60, θ_1 = 15°C, θ_2 = 40°C, α = 0.000012, ΔL = ?
Note that ΔL = the change in length, i.e. how much the iron rod will expand.

∴ $$\alpha = \frac{\Delta L}{L_1 \Delta \theta}$$

$$0.000012 = \frac{\Delta L}{60 \times (\theta_2 - \theta_1)}$$

$$0.000012 = \frac{\Delta L}{60 \times (40 - 15)}$$

$$0.000012 = \frac{\Delta L}{60 \times 25}$$

Cross multiply to obtain:

$\Delta L = 0.000012 \times 60 \times 25$

$\Delta L = 0.018m$

The iron rod will expand by 0.018m.

2. 5m of brass increases by 0.8cm when its temperature rises by 90°C. What is the linear expansivity of brass?

Solution

The two units of the lengths provided are not the same, so convert 0.8cm to metre as follows: $\frac{0.8}{100} = 0.008m$

Therefore, $\Delta L = 0.008$, $L_1 = 5$, $\Delta\theta = 90$, $\alpha = ?$

$$\alpha = \frac{\Delta L}{L_1 \Delta\theta}$$

$$\alpha = \frac{0.008}{5 \times 90}$$

$\alpha = 0.000018$

The linear expansivity of brass is 0.000018 K^{-1}.

3. An iron rod is 80cm at 120°C. What will be its length at 30°C.? (Linear expansivity of iron = 1×10^{-5} K^{-1})

Solution

This is a case of cooling the iron rod (i.e. contraction). Here, we are going to assume that the iron rod was heated from 30°C to 120°C.

So, the length at 30°C is L_1, while the length at 120°C is L_2

Therefore, $L_2 = 80$, $\theta_1 = 30°C$, $\theta_2 = 120°C$, $\alpha = 1 \times 10^{-5}$, $L_1 = ?$

$$\therefore \alpha = \frac{L_2 - L_1}{L_1(\theta_2 - \theta_1)}$$

$$1 \times 10^{-5} = \frac{80 - L_1}{L_1(120 - 30)}$$

$$1 \times 10^{-5} = \frac{80 - L_1}{L_1(90)}$$

$$0.00001 = \frac{80 - L_1}{90L_1} \quad \text{(Note that } 1 \times 10^{-5} = 1/10^5 = \frac{1}{100000} = 0.00001\text{)}$$

Cross multiply to obtain:

$$0.00001 \times 90L_1 = 80 - L_1$$

$$0.0009 L_1 = 80 - L_1$$

$$0.0009 L_1 + L_1 = 80$$

$$1.0009 L_1 = 80 \quad \text{(Note that } L_1 \text{ also means } 1L_1\text{)}$$

$$L_1 = \frac{80}{1.0009}$$

$$L_1 = 79.93$$

Its length at 30°C will be 79.93cm

4. 105cm of brass increases by 0.21cm when heated. Calculate the final temperature of the brass if it was originally at 24°C. (Take linear expansivity of brass as 1.9×10^{-5} K^{-1})

Solution
Given in the question are: $L_1 = 105$, $\theta_1 = 24°C$, $\alpha = 1.9 \times 10^{-5}$, $\Delta L = 0.21$, $\theta_2 = ?$

$$\therefore \alpha = \frac{\Delta L}{L_1 \Delta \theta}$$

$$1.9 \times 10^{-5} = \frac{0.21}{105 \times (\theta_2 - \theta_1)}$$

$$0.000019 = \frac{0.21}{105 \times (\theta_2 - 24)} \quad \text{(Note that } 1.9 \times 10^{-5} = 1.9/10^5 = \frac{1.9}{100000} = 0.000019\text{)}$$

$$0.000019 = \frac{0.21}{(105\theta_2 - 2520)}$$

Cross multiply to obtain:

$$0.000019(105\theta_2 - 2520) = 0.21$$

$$0.001995\theta_2 - 0.04788 = 0.21$$

$$0.001995\theta_2 = 0.21 + 0.04788$$

$$0.001995\theta_2 = 0.25788$$

$$\theta_2 = \frac{0.25788}{0.001995}$$

$$\theta_2 = 129$$

The final temperature of the brass is 129°C

Another method that can be used to solve this question is to calculate the change in temperature, Δθ. This is calculated as follows:

$$\alpha = \frac{\Delta L}{L_1 \Delta \theta}$$

$$1.9 \times 10^{-5} = \frac{0.21}{105 \times \Delta\theta}$$

$$0.000019 = \frac{0.21}{105 \Delta\theta}$$

Cross multiply to obtain:

$$0.000019 \times 105\Delta\theta = 0.21$$

$$0.001995\Delta\theta = 0.21$$

$$\Delta\theta = \frac{0.21}{0.001995}$$

$$\Delta\theta = 105$$

But, $\Delta\theta = \theta_2 - \theta_1$

$$105 = \theta_2 - 24$$

$$105 + 24 = \theta_2$$

$$\theta_2 = 129 \quad \text{(As obtained before)}$$

∴ The final temperature of the brass is 129°C

5. A metal of length 2m is heated to a length of 2.0016m. Through what temperature was it heated if the linear expansivity of the metal is 1.7×10^{-5} K^{-1}.

Solution

Given: $L_1 = 2$, $L_2 = 2.0016$, $\alpha = 1.7 \times 10^{-5}$, $\Delta\theta = ?$

$$\alpha = \frac{L_2 - L_1}{L_1 \Delta\theta}$$

$$1.7 \times 10^{-5} = \frac{2.0016 - 2}{2 \times \Delta\theta}$$

$$0.000017 = \frac{0.0016}{2\Delta\theta}$$

Cross multiply to obtain:

$$0.000017 \times 2\Delta\theta = 0.0016$$

$$0.000034\Delta\theta = 0.0016$$

$$\Delta\theta = \frac{0.0016}{0.000034}$$

$$\Delta\theta = 47$$

The metal was heated through 47°C

6. A rail line is to be constructed with steel bars, each of length 3.5m at 25°C. What is the safety gap that must be left between any two bars in order to allow for expansion, if the expected maximum temperature of the day is 42°C? (Linear expansivity of steel = 1.1×10^{-5} K^{-1})

Solution

The safety gap between any two bars is the amount by which each bar will expand. This is equal to the change in length of each bar.

Given in the question are: L_1 = 3.5, θ_1 = 25°C, θ_2 = 42°C, α = 1.1×10^{-5}, ΔL = ?

$$\therefore \alpha = \frac{\Delta L}{L_1 \Delta \theta}$$

$$1.1 \times 10^{-5} = \frac{\Delta L}{3.5 \times (\theta_2 - \theta_1)}$$

$$0.000011 = \frac{\Delta L}{3.5 \times (42 - 25)}$$

$$0.000011 = \frac{\Delta L}{3.5 \times 17}$$

Cross multiply to obtain:

ΔL = 0.000011 x 3.5 x 17

ΔL = 0.000655

The safety gap needed between any two bars is 0.000655m.

7. The ratio of the linear expansivity of copper to that of iron is approximately 1:4. An iron material and a copper material of the same shape expand by the same amount per unit rise in temperature. What is the ratio of the length of iron to that of copper?

Solution

Let the linear expansivity of copper be α. Therefore the linear expansivity of iron will be = 4α (Since ratio of expansivity of copper to that of iron is 1:4)

For copper: $\alpha = \dfrac{\Delta L}{L_c \Delta\theta}$ where L_c is the length of copper

$$\alpha = \dfrac{\Delta L}{L_c \times 1}$$ (Δθ =1, from the statement "per unit rise in temperature")

∴   ΔL = αL_c ……………………. Equation 1

Similarly, the change in length of iron will be given by:

ΔL = 4αL_i …………………….. Equation 2   (Where L_i is the length of iron)

Since the iron and copper both expand by the same amount, then equation 1 will be equal to equation 2. Equating only the right hand sides of both equations gives:

4αL_i = αL_c

Divide both sides of the equation by α. This gives:

4L_i = L_c

Divide both sides of the equation by 4L_c in order to obtain the ration L_i/L_c. This gives:

$$\dfrac{4L_i}{4L_c} = \dfrac{L_c}{4L_c}$$

$$\dfrac{L_i}{L_c} = \dfrac{1}{4}$$

Therefore, the ratio of the length of iron to that of copper is 1:4.

Exercise 11

1. A brass rod has a length of 12m when the temperature is 15°C. By how much will it expand when the temperature rises to 52°C? (Linear expansivity of brass = 0.000019 K^{-1})

2. 9m of brass increases by 1.5cm when its temperature rises by 88°C. What is the linear expansivity of brass?

3. An iron rod is 220cm at 98°C. What will be its length at 42°C.? (Linear expansivity of iron = 1×10^{-5} K^{-1})

4. 2.4m of brass increases by 0.005m when heated. Calculate the final temperature of the brass if it was originally at 21°C. (Take linear expansivity of brass as 1.8×10^{-5} K^{-1})

5. A metal of length 120cm is heated to a length of 122cm. Through what temperature was it heated if the linear expansivity of the metal is 1.1×10^{-5} K^{-1}.

6. A bridge is to be constructed with steel bars, each of length 6m at 20°C. What is the safety gap that must be left between any two bars if the expected maximum temperature of the day is 30°C? (Linear expansivity of steel = 1.2×10^{-5} K^{-1})

7. The ratio of the linear expansivity of iron to that of brass is approximately 2:3. An iron material and a brass material of the same shape expand by the same amount per unit rise in temperature. What is the ratio of the length of iron to that of brass?

CHAPTER 12
AREA EXPANSIVITY

The change in area of an object can be expressed in terms of a quantity called area expansivity. Area expansivity can be expressed as follows:

$$\beta = \frac{\Delta A}{A_1 \Delta \theta}$$

Or, $$\beta = \frac{A_2 - A_1}{A_1(\theta_2 - \theta_1)}$$

Where β = area expansivity, $\Delta A = (A_2 - A_1)$ = change in area, A_1 = the initial area, A_2 = final area. A_2 is greater than A_1. Also, $\Delta\theta = (\theta_2 - \theta_1)$ = change in temperature, and θ_2 is a higher temperature than θ_1.

The unit of area expansivity is K^{-1} or /°C. Note that area expansivity is also called superficial expansivity.

Area expansivity and linear expansivity are related by the expression:

$$\beta = 2\alpha$$

This means that when linear expansivity of a body is known, the area expansivity of the body can be determined by multiplying the value of the linear expansivity by two.

Examples

1. A metal has an area of 500mm² when the temperature is 10°C. By how much will the area increase when the temperature rises to 30°C? (Area expansivity of the metal = 0.000025 K^{-1})

Solution

Given in the question are: A_1 = 500, θ_1 = 10°C, θ_2 = 30°C, β = 0.000025, ΔA = ? Note that ΔA = the increase in area of the metal.

$$\therefore \quad \beta = \frac{\Delta A}{A_1 \Delta \theta}$$

$$0.000025 = \frac{\Delta A}{500 \times (\theta_2 - \theta_1)}$$

$$0.000025 = \frac{\Delta A}{500 \times (30 - 10)}$$

$$0.000025 = \frac{\Delta A}{500 \times 20}$$

Cross multiply to obtain:

$\Delta A = 0.000025 \times 500 \times 20$

$\Delta A = 0.25$

The metal will increase in area by $0.25 mm^2$.

2. $8m^2$ of iron increases by $84cm^2$ when its temperature rises by 45°C. What is the Area expansivity of iron?

Solution

The two units of the Area provided are not the same, so convert $84cm^2$ to square metre as follows: $\frac{84}{100 \times 100} = 0.0084 m^2$ (Note that in order to convert cm^2 to m^2, divide the value in cm^2 by 100×100)

Therefore, $\Delta A = 0.0084$, $A_1 = 8$, $\Delta \theta = 45$, $\beta = ?$

$$\beta = \frac{\Delta A}{A_1 \Delta \theta}$$

$$\beta = \frac{0.0084}{8 \times 45}$$

$$\beta = \frac{0.0084}{360}$$

$$\beta = 0.000023$$

The Area expansivity of the iron is 0.000023 K^{-1}.

3. An iron rod has an area of 6m^2 at 80°C. What will be its area at 25°C.? (Linear expansivity of iron = 1.2 x 10^{-5} K^{-1})

Solution

Here, the assumption will be that the iron rod was heated from 25°C to 80°C.

So, the area at 25°C is A_1, while the Area at 80°C is A_2

Therefore, A_2 = 6, θ_1 = 25°C, θ_2 = 80°C, β = 2α = 2 x 1.2 x 10^{-5} = 2.4 x 10^{-5}, A_1 = ?

Note that the linear expansivity given has been converted to area expansivity by multiplying it by 2.

$$\therefore \quad \beta = \frac{A_2 - A_1}{A_1(\theta_2 - \theta_1)}$$

$$2.4 \times 10^{-5} = \frac{6 - A_1}{A_1(80 - 25)}$$

$$2.4 \times 10^{-5} = \frac{6 - A_1}{A_1(55)}$$

$$0.000024 = \frac{6 - A_1}{55A_1} \quad \text{(Note that } 2.4 \times 10^{-5} = 2.4/10^5 = \frac{2.4}{100000} = 0.000024\text{)}$$

Cross multiply to obtain:

0.000024 x 55A_1 = 6 – A_1

0.00132A_1 = 6 – A_1

0.00132A_1 + A_1 = 6

1.00132A_1 = 6   (Note that A_1 also means 1A_1)

$$A_1 = \frac{6}{1.00132}$$

$A_1 = 5.99$

Its area at 25°C will be 5.99m^2

4. 0.094cm^2 of brass increases in area by 0.00032cm^2 when heated. Calculate the final temperature of the brass if it was originally at 41°C. (Take linear expansivity of brass as 1.8×10^{-5} K^{-1})

Solution

Let us first determine the change in temperature of the brass.

Given in the question are: $A_1 = 0.094$, $\theta_1 = 41°C$, $\beta = 2\alpha = 2 \times 1.8 \times 10^{-5} = 3.6 \times 10^{-5}$, $\Delta A = 0.00032$, $\Delta\theta = ?$

$\therefore \quad \beta = \dfrac{\Delta A}{A_1 \Delta\theta}$

$3.6 \times 10^{-5} = \dfrac{0.00032}{0.094 \times \Delta\theta}$

$0.000036 = \dfrac{0.00032}{0.094 \Delta\theta}$

Cross multiply to obtain:

$0.000036 \times 0.094\Delta\theta = 0.00032$

$0.000003384\Delta\theta = 0.00032$

$\Delta\theta = \dfrac{0.00032}{0.000003384}$

$\Delta\theta = 94.6$

But, $\Delta\theta = \theta_2 - \theta_1$

$94.6 = \theta_2 - 41$

$96.6 + 41 = \theta_2$

$\theta_2 = 135.6$

∴ The final temperature of the brass is 135.6°C

5. 20m² of brass increases in area by 0.028m² when its temperature rises by 39°C. What is the linear expansivity of brass?

Solution

Given: $\Delta A = 0.028$, $A_1 = 20$, $\Delta\theta = 39$, $\beta = ?$

$$\beta = \frac{\Delta A}{A_1 \Delta\theta}$$

$$\beta = \frac{0.028}{20 \times 39}$$

$$\beta = \frac{0.028}{780}$$

$$\beta = 0.000036$$

But, $\beta = 2\alpha$

∴ $\alpha = \frac{\beta}{2}$

$$\alpha = \frac{0.000036}{2}$$

$$\alpha = 0.000018$$

The linear expansivity of brass is $0.000018 \, K^{-1}$

6. A plane metal sheet in the shape of a rectangle has a length of 50cm and a width of 20cm. If it is heated from 24°C to 46°C, calculate the increase in area of the metal. (Linear expansivity of the metal = $1.4 \times 10^{-5} \, K^{-1}$)

Solution

Area of rectangle = length x width

$$= 50 \times 20 = 1000 \text{cm}^2$$

Therefore, $A_1 = 1000$, $\theta_1 = 24°C$, $\theta_2 = 46°C$, $\beta = 2\alpha = 2 \times 1.4 \times 10^{-5} = 2.8 \times 10^{-5}$, $\Delta A = ?$

$$\therefore \quad \beta = \frac{\Delta A}{A_1 \Delta \theta}$$

$$2.8 \times 10^{-5} = \frac{\Delta A}{1000 \times (\theta_2 - \theta_1)}$$

$$0.000028 = \frac{\Delta A}{1000 \times (46 - 24)}$$

$$0.000028 = \frac{\Delta A}{1000 \times 22}$$

Cross multiply to obtain:

$$\Delta A = 0.000028 \times 1000 \times 22$$

$$\Delta A = 0.616$$

The metal will increase in area by 0.616cm^2.

Exercise 12

1. An iron plate has an area of 220cm^2 when the temperature is $22°C$. By how much will the area increase when the temperature rises to $50°C$? (Area expansivity of iron = 0.000024 K^{-1})

2. 5m^2 of a metal increases by 52cm^2 when its temperature rises from $31°C$ to $75°C$. What is the Area expansivity of the metal?

3. An iron disc has an area of 1200cm^2 at $100°C$. What will be its area at $28°C$.? (Linear expansivity of iron = $1.2 \times 10^{-5} \text{ K}^{-1}$)

4. 0.66m^2 of brass increases in area by 0.00045m^2 when heated. Calculate the final temperature of the brass if it was originally at $26°C$. (Take linear expansivity of brass as $1.8 \times 10^{-5} \text{ K}^{-1}$)

5. 38m² of brass increases in area by 0.055m² when its temperature rises by 40°C. What is the linear expansivity of brass?

6. A plane metal sheet in the shape of a square has sides of length 250cm. If it is heated from 50°C to 80°C, calculate the increase in area of the metal. (Linear expansivity of the metal = 1.4×10^{-5} K^{-1})

7. 16m² of iron increases in area by 0.0098m² when heated. Calculate the final temperature of the iron if it was originally at 21°C. (Take linear expansivity of iron as 1×10^{-5} K^{-1})

CHAPTER 13
VOLUME EXPANSIVITY

The change in volume of an object can be expressed in terms of a quantity called volume expansivity. Volume expansivity can be expressed as follows:

$$\gamma = \frac{\Delta V}{V_1 \Delta \theta}$$

Or, $$\gamma = \frac{V_2 - V_1}{V_1(\theta_2 - \theta_1)}$$

Where γ = Volume expansivity, $\Delta V = (V_2 - V_1)$ = change in Volume, V_1 = the initial volume, V_2 = final volume. V_2 is greater than V_1. Also, $\Delta \theta = (\theta_2 - \theta_1)$ = change in temperature, and θ_2 is a higher temperature than θ_1.

Note that volume expansivity is also called cubic expansivity.

Volume expansivity and linear expansivity are related by the expression:

$$\gamma = 3\alpha$$

This means that when linear expansivity of a body is known, the volume expansivity of the body can be determined by multiplying the value of the linear expansivity by three.

Examples

1. The area expansivity of a metal is $2.6 \times 10^{-5} K^{-1}$. What is the volume expansivity of the metal?

Solution

Recall that: $\beta = 2\alpha$

$\therefore \quad \alpha = \frac{\beta}{2}$

$$\alpha = \frac{0.000026}{2}$$

$$\alpha = 0.000013$$

But, $\gamma = 3\alpha$

$$\gamma = 3 \times 0.000013$$

$$\gamma = 0.000039$$

The volume expansivity of the metal is 0.000039 K^{-1}.

2. An iron box has a volume of 24m^2 when the temperature is 21°C. By how much will the volume increase when the temperature rises to 50°C? (Volume expansivity of the iron = 0.000038 K^{-1})

Solution

Given in the question are: V_1 = 24, θ_1 =21°C, θ_2 = 50°C, γ = 0.000038, ΔV = ?
Note that ΔV = the increase in volume of the iron.

$$\therefore \quad \gamma = \frac{\Delta V}{V_1 \Delta \theta}$$

$$0.000038 = \frac{\Delta V}{24 \times (\theta_2 - \theta_1)}$$

$$0.000038 = \frac{\Delta V}{24 \times (50 - 21)}$$

$$0.000038 = \frac{\Delta V}{24 \times 29}$$

Cross multiply to obtain:

$$\Delta V = 0.000038 \times 24 \times 29$$

$$\Delta V = 0.0264$$

The iron will increase in volume by 0.0264m².

3. A solid brass material of volume 15m³ increases by 44000cm³ when its temperature rises by 52°C. What is the cubic expansivity of brass?

Solution

The two units of the volume provided are not the same, so convert 44000cm³ to cubic metre as follows: $\frac{44000}{100 \times 100 \times 100} = 0.044 m^2$ (Note that in order to convert cm³ to m³, divide the value in cm³ by 100 x 100 x 100)

Therefore, $\Delta V = 0.044$, $V_1 = 15$, $\Delta\theta = 52$, $\gamma = ?$

$$\gamma = \frac{\Delta V}{V_1 \Delta\theta}$$

$$\gamma = \frac{0.044}{15 \times 52}$$

$$\gamma = \frac{0.044}{780}$$

$$\gamma = 0.000056$$

The cubic expansivity of brass is 0.000056 K⁻¹.

4. A cylindrical brass material has a volume of 260cm³ at 119°C. What will be its volume at 60°C.? (Linear expansivity of brass = 1.7×10^{-5} K⁻¹)

Solution

Assume that brass was heated from 60°C to 119°C.

Therefore, the volume at 60°C is V_1, while the Volume at 119°C is V_2

So, $V_2 = 260$, $\theta_1 = 60°C$, $\theta_2 = 119°C$, $\gamma = 3\alpha = 3 \times 1.7 \times 10^{-5} = 5.1 \times 10^{-5}$, $V_1 = ?$

Note that the linear expansivity given has been converted to volume expansivity by multiplying it by 3.

$$\therefore \quad \gamma = \frac{V_2 - V_1}{V_1(\theta_2 - \theta_1)}$$

$$5.1 \times 10^{-5} = \frac{260 - V_1}{V_1(119 - 60)}$$

$$5.1 \times 10^{-5} = \frac{260 - V_1}{V_1(59)}$$

$$0.000051 = \frac{260 - V_1}{59V_1}$$

Cross multiply to obtain:

$$0.000051 \times 59V_1 = 260 - V_1$$

$$0.003009V_1 = 260 - V_1$$

$$0.003009V_1 + V_1 = 260$$

$$1.003009V_1 = 260$$

$$V_1 = \frac{260}{1.003009}$$

$$V_1 = 259.2$$

Its volume at 60°C will be 259.2m^3

5. 108cm^3 of iron increases in volume by 0.82cm^3 when heated. Calculate the final temperature of the iron if it was originally at 29°C. (Take linear expansivity of iron as 1.1×10^{-5} K^{-1})

Solution

Let us first determine the change in temperature of the iron.

Given in the question are: $V_1 = 108$, $\theta_1 = 29°C$, $\gamma = 3\alpha = 3 \times 1.1 \times 10^{-5} = 3.3 \times 10^{-5}$, $\Delta V = 0.82$, $\Delta\theta = ?$

$$\therefore \gamma = \frac{\Delta V}{V_1 \Delta\theta}$$

$$3.3 \times 10^{-5} = \frac{0.82}{108 \times \Delta\theta}$$

$$0.000033 = \frac{0.82}{108\Delta\theta}$$

Cross multiply to obtain:

$0.000033 \times 108\Delta\theta = 0.82$

$0.003564\Delta\theta = 0.82$

$$\Delta\theta = \frac{0.82}{0.003564}$$

$\Delta\theta = 230$

But, $\Delta\theta = \theta_2 - \theta_1$

$230 = \theta_2 - 29$

$230 + 29 = \theta_2$

$\theta_2 = 259$

∴ The final temperature of the iron is 259°C

6. 1600mm³ of brass increases in volume by 3.7mm³ when its temperature rises by 43°C. What is the linear expansivity of brass?

Solution
Given: $\Delta V = 3.7$, $V_1 = 1600$, $\Delta\theta = 43$, $\gamma = ?$

$$\gamma = \frac{\Delta V}{V_1 \Delta\theta}$$

$$\gamma = \frac{3.7}{1600 \times 43}$$

$$\gamma = \frac{3.7}{68800}$$

$$\gamma = 0.0000538$$

But, $\gamma = 3\alpha$

$$\therefore \quad \alpha = \frac{\gamma}{3}$$

$$\alpha = \frac{0.0000538}{3}$$

$$\alpha = 0.000018$$

The linear expansivity of brass is 0.000018 K^{-1}

7. A solid metal cube of side 12cm is heated from 20°C to 80°C. If the linear expansivity of the metal is 1.2×10^{-5} K^{-1}, calculate the increase in volume of the cube.

Solution

Volume of cube = length x length x length

$$= 12 \times 12 \times 12 = 1728 cm^3$$

Therefore, $V_1 = 1728$, $\theta_1 = 20°C$, $\theta_2 = 80°C$, $\gamma = 3\alpha = 3 \times 1.2 \times 10^{-5} = 3.6 \times 10^{-5}$, $\Delta V = ?$

$$\therefore \quad \gamma = \frac{\Delta V}{V_1 \Delta \theta}$$

$$3.6 \times 10^{-5} = \frac{\Delta V}{1728 \times (\theta_2 - \theta_1)}$$

$$0.000036 = \frac{\Delta V}{1728 \times (80 - 20)}$$

$$0.000036 = \frac{\Delta V}{1728 \times 60}$$

Cross multiply to obtain:

$\Delta V = 0.000036 \times 1728 \times 60$

$\Delta V = 3.73$

The increase in volume of the cube is $3.73 cm^3$.

Exercise 13

1. The volume expansivity of a metal is $4 \times 10^{-5} K^{-1}$. What is the area expansivity of the metal?

2. An iron box has a volume of $30 m^3$ when the temperature is 18°C. By how much will the volume increase when the temperature rises to 64°C? (Volume expansivity of the iron = $0.000038 K^{-1}$)

3. A solid brass material of volume $27 m^3$ increases by $86500 cm^3$ when its temperature rises by 49°C. What is the cubic expansivity of brass?

4. A cylindrical brass material has a volume of $1950 mm^3$ at 104°C. What will be its volume at 32°C.? (Linear expansivity of brass = $1.8 \times 10^{-5} K^{-1}$)

5. $3.52 m^3$ of iron increases in volume by $0.002 m^3$ when heated. Calculate the final temperature of the iron if it was originally at 25°C. (Take linear expansivity of iron as $1 \times 10^{-5} K^{-1}$)

6. $3400 mm^3$ of brass increases in volume by $7.8 mm^3$ when its temperature rises by 42°C. What is the linear expansivity of brass?

7. A solid metal cube of side 100mm is heated from 15°C to 100°C. If the linear expansivity of the metal is $1.4 \times 10^{-5} K^{-1}$, calculate the increase in volume of the cube.

CHAPTER 14
REAL AND APPARENT CUBIC EXPANSIVITY

Real cubic expansivity and apparent cubic expansivity becomes necessary when a liquid is heated in a container. The real cubic expansivity involves the actual increase in the volume of a liquid. The apparent cubic expansivity involves the increase in the volume of a liquid when it is heated in an expansible vessel. In chapter 13, the cubic expansivity of solid was discussed. For liquid the cubic expansivity is called apparent cubic expansivity since a liquid is usually heated in a vessel/container. Similar to cubic expansivity of solids, the apparent cubic expansivity of a liquid is given by:

$$\gamma_a = \frac{\Delta V}{V_1 \Delta \theta}$$

Or, $$\gamma_a = \frac{V_2 - V_1}{V_1(\theta_2 - \theta_1)}$$

Where γ_a = apparent cubic expansivity, while other symbols have their usual meanings. Note that ΔV is usually regarded as the apparent change in volume of the liquid.

In terms of mass or volume of liquid expelled when a liquid is heated in a vessel, the apparent cubic expansivity of a liquid can be expressed as follows:

$$\gamma_a = \frac{\text{mass of liquid expelled}}{\text{mass of liquid remaining} \times \text{temperature change}}$$

Or, $$\gamma_a = \frac{\text{volume of liquid expelled}}{\text{volume of liquid left} \times \text{temperature change}}$$

Real cubic expansivity, apparent cubic expansivity and the cubic expansivity of the container carrying the liquid are related by:

$$\gamma_r = \gamma_a + \gamma$$

Where γ_r = real cubic expansivity, γ_a = apparent cubic expansivity and γ = the cubic expansivity of the container carrying the liquid.

Examples

1. A liquid in a container has a volume of 2m³. When it is heated through a temperature of 100°C, it volume increased by 0.01m³. If the volume expansivity of the container is 0.0002K⁻¹, calculate:

a. the apparent expansivity of the liquid

b. the real expansivity of the liquid

Solutions

a. $\gamma_a = \dfrac{\Delta V}{V_1 \Delta \theta}$

$\gamma_a = \dfrac{0.01}{2 \times 100}$

$\gamma_a = \dfrac{0.01}{200}$

$\gamma_a = 0.00005$

The apparent expansivity of the liquid is $0.00005 K^{-1}$ or $5 \times 10^{-5} K^{-1}$

b. $\gamma_r = \gamma_a + \gamma$

$\gamma_r = 0.00005 + 0.0002$

$\gamma_r = 0.00025$

The real expansivity of the liquid is $0.00025 K^{-1}$ or $2.5 \times 10^{-4} K^{-1}$

2. The volume of the bulb of a mercury-in-glass thermometer is 0.6cm³. The temperature of the mercury increases from 0°C to 100°C. If the cross-sectional area of the tube is $1.5 \times 10^{-4} cm^2$, calculate:

a. the apparent increase in the volume of the mercury

b. the distance between the fixed points. ($\gamma_a = 1.4 \times 10^{-5} K^{-1}$)

Solutions

a. $\gamma_a = \dfrac{\Delta V}{V_1 \Delta \theta}$

$1.4 \times 10^{-5} = \dfrac{\Delta V}{0.6 \times (\theta_2 - \theta_1)}$ (Note that the mercury is contained in the tube and it is at the 0°C mark when the temperature is 0°C. So, the volume of the tube, i.e. 0.6cm³ is also the volume of the mercury when it is at 0°C)

$0.000014 = \dfrac{\Delta V}{0.6 \times (100 - 0)}$

$0.000014 = \dfrac{\Delta V}{0.6 \times 100}$

Cross multiply to obtain:

$\Delta V = 0.000014 \times 0.6 \times 100$

$\Delta V = 0.00084$

The apparent increase in volume of mercury is 0.00084cm³.

b. Let d be the distance between the fixed points. This distance represents the height of the tube from 0°C to 100°C.

Recall that: Volume = cross-sectional area x height

∴ $0.00084 = 1.5 \times 10^{-4} \times d$

$0.00084 = 0.00015d$

$d = \dfrac{0.00084}{0.00015}$

$d = 5.6$

The distance between the fixed points is 5.6cm.

3. A liquid is heated in a container. If the apparent cubic expansivity of the liquid is $1.2 \times 10^{-4} K^{-1}$, and the linear expansivity of the container is $1.8 \times 10^{-5} K^{-1}$, calculate the real cubic expansivity of the liquid.

Solution

Linear expansivity of the container, $\alpha = 1.8 \times 10^{-5} K^{-1}$

But, $\gamma = 3\alpha$

∴ Cubic expansivity of the container, $\gamma = 3 \times 1.8 \times 10^{-5} = 5.4 \times 10^{-5} K^{-1}$

Also, $\gamma_r = \gamma_a + \gamma$

$\gamma_r = 1.2 \times 10^{-4} + 5.4 \times 10^{-5}$

$\gamma_r = 0.00012 + 0.000054$

$\gamma_r = 0.000174$

The real cubic expansivity is $0.000174 K^{-1}$ or $1.74 \times 10^{-4} K^{-1}$

Note that from the above calculation, 1.2×10^{-4} and 5.4×10^{-5} can only be added directly if they are both converted to have the same power on 10. Applying this method will make the addition to be solved as either:

$12 \times 10^{-5} + 5.4 \times 10^{-5} = (12 + 5.4) \times 10^{-5} = 17.4 \times 10^{-5} = 1.74 \times 10^{-4}$ (Note that 1.2×10^{-4} has been converted to 12×10^{-5})

Or, $1.2 \times 10^{-4} + 0.54 \times 10^{-4} = (1.2 + 0.54) \times 10^{-4} = 1.74 \times 10^{-4}$ (Note that 5.4×10^{-5} has been converted to 0.54×10^{-4})

This method is usually very helpful when the power of 10 is very high. However, when the power of 10 is not high, then the expanded method used in the example above (i.e. 0.00012 + 0.000054) will be easier to use. It is very helpful if the topic 'standard form' in mathematics is well understood.

4. A glass bottle full of mercury has a mass of 200g. When the bottle is heated through 42°C, 0.94g of mercury is expelled. Calculate the mass of mercury remaining in the bottle. (Real cubic expansivity of mercury is $1.7 \times 10^{-4} K^{-1}$, linear expansivity of glass is $8.5 \times 10^{-6} K^{-1}$)

Solution

We have to first determine the apparent expansivity of mercury.

Linear expansivity of the glass, $\alpha = 8.5 \times 10^{-6}$

But, $\gamma = 3\alpha$

∴ Cubic expansivity of the glass, $\gamma = 3 \times 8.5 \times 10^{-6} = 25.5 \times 10^{-6}$

$$\gamma_r = \gamma_a + \gamma$$

$$1.7 \times 10^{-4} = \gamma_a + 25.5 \times 10^{-6}$$

$$0.00017 = \gamma_a + 0.0000255$$

$$\gamma_a = 0.00017 - 0.0000255$$

$$\gamma_a = 0.000145$$

But, $\gamma_a = \dfrac{\text{mass of liquid expelled}}{\text{mass of liquid remaining} \times \text{temperature change}}$

From the question: mass of liquid expelled = 0.94, temperature change = 42, mass of liquid remaining = ?

$$\gamma_a = \dfrac{\text{mass of liquid expelled}}{\text{mass of liquid remaining} \times \text{temperature change}}$$

$$0.000145 = \dfrac{0.94}{m \times 42} \quad \text{(where m is the mass remaining)}$$

Cross multiply to obtain:

$0.000145 \times 42 \times m = 0.94$

$$0.00609m = 0.94$$

$$\therefore \quad m = \frac{0.94}{0.00609}$$

$$m = 154.4$$

The mass of mercury remaining in the bottle is 154.4g

5. A bottle full of mercury has a mass of 350g. When the bottle is heated from 22°C to 58°C, 1.6g of mercury is expelled. Calculate

a. the mass of mercury left in the bottle

b. the mass of the bottle.

(Real cubic expansivity of mercury is $1.8 \times 10^{-4} K^{-1}$, linear expansivity of the bottle is $8 \times 10^{-6} K^{-1}$).

Solution

a. Let us first determine the apparent expansivity of mercury.

Linear expansivity of the bottle, $\alpha = 8 \times 10^{-6}$

But, $\gamma = 3\alpha$

$\therefore$ Cubic expansivity of the bottle, $\gamma = 3 \times 8 \times 10^{-6} = 24 \times 10^{-6}$

$$\gamma_r = \gamma_a + \gamma$$

$$1.8 \times 10^{-4} = \gamma_a + 24 \times 10^{-6}$$

$$0.00018 = \gamma_a + 0.000024$$

$$\gamma_a = 0.00018 - 0.000024$$

$$\gamma_a = 0.000156$$

But, $\gamma_a = \dfrac{\text{mass of liquid expelled}}{\text{mass of liquid left} \times \text{temperature change}}$

From the question: mass of liquid expelled = 1.6, temperature change = 58 – 22 = 36, mass of liquid remaining = ?

$$\gamma_a = \dfrac{\text{mass of liquid expelled}}{\text{mass of liquid left} \times \text{temperature change}}$$

$0.000156 = \dfrac{1.6}{m \times 36}$ (where m is the mass of mercury left)

Cross multiply to obtain:

0.000156 x 36 x m = 1.6

0.005616m = 1.6

∴ $m = \dfrac{1.6}{0.005616}$

m = 284.9

The mass of mercury left in the bottle is 284.9g

b. Mass of mercury left + mass of mercury expelled + mass of bottle = mass of bottle and content.

284.9 + 1.6 + x = 350   (where x is the mass of the bottle)

286.5 + x = 350

x = 350 – 286.5

x = 63.5

The mass of the bottle is 63.5g

6. A beaker of volume 250cm³ is full of mercury. When the beaker is heated through 28°C, 1.02cm³ of mercury is expelled. Calculate the apparent expansivity of mercury.

Solution

From the question: volume of liquid expelled = 1.02, volume of liquid remaining = 250 − 1.02 = 248.98, temperature change = 28, apparent cubic expansivity of mercury, γ_a = ?

$$\gamma_a = \frac{\text{volume of liquid expelled}}{\text{volume of liquid remaining} \times \text{temperature change}}$$

$$\gamma_a = \frac{1.02}{248.98 \times 28}$$

$$\gamma_a = \frac{1.02}{6971.44}$$

$$\gamma_a = 0.000146$$

The apparent expansivity of mercury is $0.000146 K^{-1}$ or $1.46 \times 10^{-4} K^{-1}$

7. In an experiment to determine the apparent expansivity of a liquid in the laboratory, the following information was obtained.

Mass of empty bottle = 12g
Mass of empty bottle + liquid = 48g
Mass of empty bottle + liquid left = 47.64g
Temperature of liquid before heating = 25°C
Temperature of liquid after heating = 59°C

From the information given above, calculate the apparent expansivity of the liquid.

Solution
Mass of liquid expelled = (Mass of empty bottle + liquid) − (Mass of empty bottle + liquid left)

Mass of liquid expelled = 48 − 47.64 = 0.36g

Mass of liquid left = (Mass of empty bottle + liquid left) − (Mass of empty bottle)

Mass of liquid left = 47.64 − 12 = 35.64g

Temperature change = 59 − 25 = 34°C

$$\gamma_a = \frac{\text{mass of liquid expelled}}{\text{mass of liquid left} \times \text{temperature change}}$$

$$\gamma_a = \frac{0.36}{35.64 \times 34}$$

$$\gamma_a = \frac{0.36}{1211.76}$$

$$\gamma_a = 0.000297$$

The apparent expansivity of the liquid is $0.000297 K^{-1}$ or $2.97 \times 10^{-4} K^{-1}$

Exercise 14

1. A liquid in a container has a volume of $3.6 m^3$. When it is heated through a temperature of 94°C, its volume increased by $0.19 m^3$. If the volume expansivity of the container is $0.000022 K^{-1}$, calculate:

a. the apparent expansivity of the liquid

b. the real expansivity of the liquid

2. The volume of the bulb of a mercury-in-glass thermometer is $0.58 cm^3$. The temperature of the mercury increases from 0°C to 100°C. If the cross-sectional area of the tube is $1.4 \times 10^{-4} cm^2$, calculate:

a. the apparent increase in the volume of the mercury

b. the distance between the fixed points. ($\gamma_a = 1.5 \times 10^{-5} K^{-1}$)

3. A liquid is heated in a container. If the apparent cubic expansivity of the liquid is $1.6 \times 10^{-4} K^{-1}$, and the linear expansivity of the container is $1.2 \times 10^{-5} K^{-1}$, calculate the real cubic expansivity of the liquid.

4. A glass bottle full of mercury has a mass of 420g. When the bottle is heated through 40°C, 1.9g of mercury is expelled. Calculate the mass of mercury remaining in the bottle. (Real cubic expansivity of mercury is $1.8 \times 10^{-4} K^{-1}$, linear expansivity of glass is $8 \times 10^{-6} K^{-1}$)

5. A bottle full of mercury has a mass of 180g. When the bottle is heated from 30°C to 65°C, 0.9g of mercury is expelled. Calculate:

a. the mass of mercury left in the bottle

b. the mass of the bottle.

(Real cubic expansivity of mercury is $1.8 \times 10^{-4} K^{-1}$, linear expansivity of the bottle is $8.5 \times 10^{-6} K^{-1}$).

6. A beaker of volume $50 cm^3$ is full of mercury. When the beaker is heated through 31°C, $0.24 cm^3$ of mercury is expelled. Calculate the apparent expansivity of mercury.

7. In an experiment to determine the apparent expansivity of a liquid in the laboratory, the following information was obtained.

Mass of empty bottle = 10.5g
Mass of empty bottle + liquid = 46.4g
Mass of empty bottle + liquid left = 46.02g
Temperature of liquid before heating = 20°C
Temperature of liquid after heating = 55°C

From the information given above, calculate the apparent expansivity of the liquid.

CHAPTER 15
MEASUREMENT OF TEMPERATURE

The three most common temperature scales are given below.

1. **Celsius scale:** This scale has an ice point of 0°C and a steam point of 100°C. The interval between them is called the fundamental interval.

2. **Fahrenheit scale**: This scale has an ice point of 32°F and a steam point of 212°F. It has a fundamental interval of 212 – 32 = 180.

3. **Thermodynamic (Kelvin) scale**: This has an ice point of 273k and a steam point of 373k.

Note that the steam point is also called the upper fixed point while the ice point is also called the lower fixed point.

Examples

1. The ice and steam point of an ungraduated mercury thermometer are found to be 19.2cm apart. What is the temperature in °C when the length of the mercury thread above the ice point is 6.72cm?

Solution

Proportion method will be used in carrying out this calculation. The expression below summarizes this method.

$$\frac{\text{Steam point} - \text{Ice point for first scale}}{\text{Middle value} - \text{Ice point for first scale}} = \frac{\text{Steam point} - \text{Ice point for second scale}}{\text{Middle value} - \text{Ice point for second scale}}$$

This expression can easily be remembered by using the acronym expression below.

$$\frac{(S-I) \text{ for first scale}}{(M-I) \text{ for first scale}} = \frac{(S-I) \text{ for second scale}}{(M-I) \text{ for second scale}}$$

The middle value is the temperature which is to be calculated. Note that the Celsius scale is used in problems like this when the temperature in °C is to be calculated.

Let the temperature in °C be x. Also, let the first scale be the ungraduated thermometer, and the second scale be the Celsius scale. Note that the Celsius scale middle value is to be calculated.

∴ $$\frac{\text{Steam point} - \text{Ice point for first scale}}{\text{Middle value} - \text{Ice point for first scale}} = \frac{\text{Steam point} - \text{Ice point for second scale}}{\text{Middle value} - \text{Ice point for second scale}}$$

$$\frac{19.2}{6.72} = \frac{100 - 0}{x - 0}$$

$$\frac{19.2}{6.72} = \frac{100}{x}$$

$$19.2x = 100 \times 6.72$$

$$19.2x = 672$$

∴ $$x = \frac{672}{19.2}$$

$$= 35°C$$

The temperature in °C is 35°C

2. The temperature of the melting point of ice and that of steam above boiling water are marked at 40 and 120 respectively on a certain thermometer. What is:

a. the temperature in °C when the reading on this thermometer is 80?

b. the thermometer reading when the temperature is 60°C?

Solution

a. Let the temperature in °C be x. Also, let the first scale be the certain thermometer, and the second scale be the Celsius scale. The Celsius scale middle value is to be calculated.

∴ $\dfrac{\text{Steam point} - \text{Ice point for first scale}}{\text{Middle value} - \text{Ice point for first scale}} = \dfrac{\text{Steam point} - \text{Ice point for second scale}}{\text{Middle value} - \text{Ice point for second scale}}$

$$\dfrac{120-40}{80-40} = \dfrac{100-0}{x-0}$$

$$\dfrac{80}{40} = \dfrac{100}{x}$$

$$80x = 100 \times 40$$

$$80x = 4000$$

∴ $x = \dfrac{4000}{80} = 50°C$

The temperature in °C is 50°C

b. Let x be the temperature on the certain thermometer. The middle value of the certain thermometer is to be calculated.

∴ $\dfrac{\text{Steam point} - \text{Ice point for first scale}}{\text{Middle value} - \text{Ice point for first scale}} = \dfrac{\text{Steam point} - \text{Ice point for second scale}}{\text{Middle value} - \text{Ice point for second scale}}$

$$\dfrac{120-40}{x-40} = \dfrac{100-0}{60-0}$$

$$\dfrac{80}{x-40} = \dfrac{100}{60}$$

$$100(x - 40) = 80 \times 60$$

$$100x - 4000 = 4800$$

$$100x = 4800 + 4000$$

$$100x = 8800$$

$$x = \dfrac{8800}{100}$$

$$x = 88$$

The thermometer reading is 88

3. A platinum resistance thermometer has a resistance of 10.4Ω at 0°C and 14.35Ω at 100°C. Assuming that the resistance changes uniformly with temperature, calculate:

a. the temperature when the resistance is 11.19Ω

b. the resistance of the thermometer when the temperature is 45°C.

Solution

a. Let the temperature in °C be x. Also, let the first scale be the resistance thermometer, and the second scale be the Celsius scale. The Celsius scale middle value is to be calculated.

∴ $$\frac{\text{Steam point} - \text{Ice point for first scale}}{\text{Middle value} - \text{Ice point for first scale}} = \frac{\text{Steam point} - \text{Ice point for second scale}}{\text{Middle value} - \text{Ice point for second scale}}$$

$$\frac{14.35 - 10.4}{11.19 - 10.4} = \frac{100 - 0}{x - 0}$$

$$\frac{3.95}{0.79} = \frac{100}{x}$$

$3.95x = 0.79 \times 100$

$3.95x = 79$

$x = \dfrac{79}{3.95}$

$x = 20°C$

b. Let x be the resistance on the resistance thermometer. The middle value of the resistance thermometer is to be calculated.

∴ $$\frac{\text{Steam point} - \text{Ice point for first scale}}{\text{Middle value} - \text{Ice point for first scale}} = \frac{\text{Steam point} - \text{Ice point for second scale}}{\text{Middle value} - \text{Ice point for second scale}}$$

$$\frac{14.35-10-4}{x-10.4} = \frac{100-0}{45.0}$$

$$\frac{3.95}{x-10.4} = \frac{100}{45}$$

$100(x - 10.4) = 3.95 \times 45$

$100x - 1040 = 177.75$

$100x = 177.75 + 10.40$

$100x = 1217.75$

$$x = \frac{1217.75}{100}$$

$x = 12.18$

The resistance of the thermometer is 12.18Ω

4. A thermometer with an arbitrary scale of equal divisions registers -30°S at ice point and 90°S at the steam point. Calculate the Celsius temperature corresponding to 60°S.

Solution

Let the temperature in °C be x. Also, let the first scale be the arbitrary scale, and the second scale be the Celsius scale. The Celsius scale middle value is to be calculated.

∴ $$\frac{\text{Steam point} - \text{Ice point for first scale}}{\text{Middle value} - \text{Ice point for first scale}} = \frac{\text{Steam point} - \text{Ice point for second scale}}{\text{Middle value} - \text{Ice point for second scale}}$$

$$\frac{90-(-30)}{60-(-30)} = \frac{100-0}{x-0}$$

$$\frac{90+30}{60+30} = \frac{100}{x}$$

$$\frac{120}{90} = \frac{100}{x}$$

$$120x = 100 \times 90$$

$$120x = 9000$$

$$\therefore x = \frac{9000}{120}$$

$$x = 75°C$$

5. a. Convert 65°C to °F

   b. Convert 86°F to °C

Solution

a. Let the temperature in °F be x. Also, let the first scale be the Celsius scale, and the second scale be the Fahrenheit scale. The Fahrenheit scale middle value is to be calculated.

$$\therefore \frac{\text{Steam point} - \text{Ice point for first scale}}{\text{Middle value} - \text{Ice point for first scale}} = \frac{\text{Steam point} - \text{Ice point for second scale}}{\text{Middle value} - \text{Ice point for second scale}}$$

Note that the steam and ice point for the Fahrenheit scale are 212°F and 32°F respectively.

$$\therefore \frac{100 - 0}{65 - 0} = \frac{212 - 32}{x - 32}$$

$$\frac{100}{65} = \frac{180}{x - 32}$$

$$100(x - 32) = 180 \times 65$$

$$100x - 3200 = 11700$$

$$100x = 11700 + 3200$$

$$100x = 14900$$

$$\therefore \quad x = \frac{14900}{100}$$

$$x = 149°F$$

b. Let the temperature in °C be *x*. Also, let the first scale be the Fahrenheit scale, and the second scale be the Celsius scale. The Celsius scale middle value is to be calculated.

$$\therefore \quad \frac{\text{Steam point } - \text{Ice point for first scale}}{\text{Middle value } - \text{Ice point for first scale}} = \frac{\text{Steam point } - \text{Ice point for second scale}}{\text{Middle value } - \text{Ice point for second scale}}$$

$$\therefore \quad \frac{212 - 32}{86 - 32} = \frac{100 - 0}{x - 0}$$

$$\frac{180}{54} = \frac{100}{x}$$

$$180x = 100 \times 54$$

$$100x = 5400$$

$$\therefore \quad x = \frac{5400}{100}$$

$$x = 54°C$$

6. The upper fixed point of a thermometer reads 250mm while the lower fixed point reads 70mm. Calculate:

a. the room temperature if the thermometer reads 115mm at room temperature

b. the reading of the thermometer when the temperature is 60°C .

Solutions

a. Let the room temperature in °C be *x*. Also, let the first scale be the mm scale, and the second scale be the Celsius scale. The Celsius scale middle value is to be calculated.

∴ $\dfrac{\text{Steam point} - \text{Ice point for first scale}}{\text{Middle value} - \text{Ice point for first scale}} = \dfrac{\text{Steam point} - \text{Ice point for second scale}}{\text{Middle value} - \text{Ice point for second scale}}$

$$\dfrac{250-70}{115-70} = \dfrac{100-0}{x-0}$$

$$\dfrac{180}{45} = \dfrac{100}{x}$$

$180x = 100 \times 45$

$180x = 4500$

$$x = \dfrac{4500}{180}$$

$x = 25°C$

The room temperature is 25°C.

b. Let x be the reading on the thermometer. Also, let the first scale be the mm scale, and the second scale be the Celsius scale. The mm scale middle value is to be calculated.

∴ $\dfrac{\text{Steam point} - \text{Ice point for first scale}}{\text{Middle value} - \text{Ice point for first scale}} = \dfrac{\text{Steam point} - \text{Ice point for second scale}}{\text{Middle value} - \text{Ice point for second scale}}$

$$\dfrac{250-70}{x-70} = \dfrac{100-0}{60-0}$$

$$\dfrac{180}{x-70} = \dfrac{100}{60}$$

$100(x - 70) = 60 \times 180$

$100x - 7000 = 10800$

$100x = 10800 + 7000$

$100x = 17800$

$$x = \frac{17800}{100}$$

$$x = 178$$

The reading of the thermometer is 178mm

Exercise 15

1. The ice and steam point of a certain mercury thermometer are found to be 24.6cm apart. What is the temperature in °C when the length of the mercury thread above the ice point is 10cm?

2. The temperature of the melting point of ice and that of steam above boiling water are marked at 50 and 110 respectively on a certain thermometer. What is:

a. the temperature in °C when the reading on this thermometer is 60?

b. the thermometer reading when the temperature is 90°C?

3. A platinum resistance thermometer has a resistance of 12.2Ω at 0°C and 15.6Ω at 100°C. Assuming that the resistance changes uniformly with temperature, calculate:

a. the temperature when the resistance is 14.1Ω

b. the resistance of the thermometer when the temperature is 72°C.

4. A thermometer with an arbitrary scale of equal divisions registers -40°M at ice point and 120°M at the steam point. Calculate the Celsius temperature corresponding to 30°M.

5. a. Convert 45°C to °F
   b. Convert 59°F to °C

6. The upper fixed point of a thermometer reads 320mm while the lower fixed point reads 110mm. Calculate:

a. the room temperature if the thermometer reads 190mm at room temperature

b. the reading of the thermometer when the temperature is 80°C.

7. The upper fixed point of a thermometer reads 22cm while the lower fixed point reads 4cm. Calculate:

a. the room temperature if the thermometer reads 16cm at room temperature

b. the reading of the thermometer when the temperature is 75°C.

8. A platinum resistance thermometer has a resistance of 10Ω at 0°C and 14Ω at 100°C. Assuming that the resistance changes uniformly with temperature, calculate:

a. the temperature when the resistance is 12.5Ω

b. the resistance of the thermometer when the temperature is 50°C.

CHAPTER 16
HEAT ENERGY – HEAT CAPACITY AND SPECIFIC HEAT CAPACITY

The heat energy absorbed or or given out when there is a change in the temperature of a body is given by:

$$Q = mc\Delta\theta \quad \text{or} \quad Q = mc(\theta_2 - \theta_1)$$

Where Q = quantity of heat, m = mass of the object, c = specific heat capacity and $\Delta\theta = (\theta_2 - \theta_1)$ = change in temperature, and θ_2 is a higher temperature than θ_1.

Heat Capacity: The heat capacity of a body is given by:

$$C = mc$$

Where C = heat capacity, m = mass of the body in Kg and c = specific heat capacity.

The S.I unit of heat capacity is JK^{-1}, while the S.I unit of specific heat capacity is $JKg^{-1}K^{-1}$

When a body is heated electrically, then the heat absorbed by the body is given by:

$$Q = Pt \quad (\text{Where } P = IV = I^2R = V^2/R, \text{ and is the power in watt})$$

Substituting these values for P in the expression above gives:

$$Q = IVt$$

Or $\quad Q = I^2Rt$

Or $\quad Q = V^2t/R$

These heats can also be expressed as:

$$IVt = mc\Delta\theta$$

Or $\quad I^2Rt = mc\Delta\theta$

Or $\quad V^2t/R = mc\Delta\theta$

Note that when objects at different temperatures are mixed together (i.e. hot and cold objects), then the heat gained by the cold object(s) is equal to the heat lost by the hot object(s). This is done with the assumption that there is no heat lost to the surrounding.

Examples

1. How much heat is given out when a piece of iron of mass 50g and specific heat capacity $460 JKg^{-1}K^{-1}$, cools from 85°C to 25°C?

Solution

$$Q = mc\Delta\theta$$

But, $m = (\frac{50}{1000})kg = 0.05kg$

The unit of mass in the value of the specific heat capacity must be the same with the unit of the mass of the object.

∴ $Q = mc\Delta\theta$ (Note that $\Delta\theta$ is the temperature difference, i.e. the larger temperature – the lower temperature)

$= 0.05 \times 460 \times (85 - 25)$

$= 0.05 \times 460 \times 60$

$= 1380$

The amount of heat given out is 1380J

2. Hot water at a temperature of t°C is added to twice that amount of water at a temperature at 30°C. If the resulting temperature of the mixture is 50°C, calculate t.

Solution

Heat lost by Hot water = heat gained by cold water

$(mc\Delta\theta)_{HW} = (mc\Delta\theta)_{CW}$ (HW = Hot Water, and CW = Cold Water)

Let m = mass of hot water. Therefore the mass of cold water will be = 2m (Since cold water is twice the amount of hot water)

Note that hot water cools from t°C to 50°C, while cold water became heated from 30°C to 50°C. Δθ is always the larger temperature − the lower temperature.

∴   m × c × (t - 50) = 2m × c × (50 -30)

mc(t - 50) = 2mc × 20

mc(t - 50) = 40mc

Dividing both sides by mc gives:

t − 50 = 40

t = 40 + 50

t = 90°C

3. Water of mass 2kg at a temperature of 70°C is added to 500g of water at 20°C. What is the final temperature of the mixture?

Solution

Let the final temperature of the mixture be t.

The mass of the cold water = 500g = $(\frac{500}{1000})$Kg = 0.5Kg

∴ $(mc\Delta\theta)_{HW} = (mc\Delta\theta)_{CW}$

2 × c × (70 - t) = 0.5 × c × (t - 20)

Dividing both sides by c gives:

2(70-t) = 0.5(t -20)

140 - 2t = 0.5t − 10

$$140 + 10 = 0.5t + 2t$$

$$150 = 2.5t$$

$$t = \frac{150}{2.5}$$

$$t = 60°C$$

The final temperature of the mixture is 60°C

4. A man wishes to bathe with warm water at 38°C. What mass of hot water at 90°C should he add to 15kg of cold water at 24°C in order to get what he wants.

<u>Solution</u>

Heat lost by hot water = Heat gained by cold water

$$(mc\Delta\theta)_{HW} = (mc\Delta\theta)_{CW}$$

m x c x (90 - 38) = 15 x c x (38 - 24)   (m is the mass of hot water needed)

m x c x 52 = 15 x c x 14

Dividing both sides by c cancels out c to give:

m x 52 = 15 x 14

52m = 210

$$m = \frac{210}{52}$$

m = 4.04

The mass of hot water that he needs to add to the cold water is 4.04kg

5. A piece of copper of mass 40g at 200°C is transferred into a copper calorimeter of mass 60g containing 50 g of water at 25°C. Neglecting heat loses, what will be the final

temperature of the mixture? (Specific heat capacity of copper and water are 4.2Jg^{-1}K^{-1} and 0.4 Jg^{-1}K^{-1} respectively)

Solution

Heat lost by hot copper = Heat gained by calorimeter + heat gained by water

$(mc\Delta\theta)_{HC} = (mc\Delta\theta)_{Ca} + (mc\Delta\theta)_W$ (HC = hot copper, Ca = calorimeter, W = water)

All masses are in g, so there is no need to convert the masses to Kg. Let the final temperature of the mixture be t°C

∴   40 x 0.4 x (200 - t) = 60 x 0.4 x (t - 25) + 50 x 4.2 x (t - 25)   (Note that the calorimeter and its content, i.e. the water are always at the same initial and final temperature)

16(200 - t) = 24(t - 25) + 210(t - 25)

3200 - 16t = 24t - 600 + 210t - 5250

3200 + 600 + 5250 = 24t +210t + 160t

9050 = 250t

$t = \dfrac{9050}{250}$

t = 36.2°C

The final temperature of the mixture is 36.2°C

6. A piece of copper of mass 20g at 250°C is placed in a copper calorimeter of mass 50g containing 55 g of water at 30°C. Neglecting heat loses, calculate the final temperature of the mixture? (Specific heat capacity of water = 4200JKg^{-1}K^{-1}, Specific heat capacity of copper = 400 JKg^{-1}K^{-1})

Solution
Since the units of the masses of the materials in the question are not the same as that

of the masses in the specific heat capacities, we have to make them the same by converting the masses of the materials from g to Kg.

The mass of the piece of copper = 200g = $(\frac{20}{1000})$Kg = 0.02Kg

The mass of the copper calorimeter = 50g = $(\frac{50}{1000})$Kg = 0.05Kg

The mass of the water = 55g = $(\frac{55}{1000})$Kg = 0.055Kg

∴ Heat lost by hot copper = Heat gained by calorimeter + heat gained by water

$(mc\Delta\theta)_{HC} = (mc\Delta\theta)_{Ca} + (mc\Delta\theta)_{W}$ (HC = hot copper, Ca = calorimeter, W = water)

Let the final temperature of the mixture be t°C

∴ 0.02 x 400 x (250 - t) = 0.05 x 400 x (t - 30) + 0.055 x 4200 x (t - 30)   (Note that the calorimeter and its content, i.e. the water are always at the same initial and final temperature)

8(250 - t) = 20(t - 30) + 231(t - 30)

2000 - 8t = 20t - 600 + 231t - 6930

2000 + 6930 + 600 = 20t + 231t + 8t

9530 = 259t

$t = \frac{9530}{259}$

t = 36.8°C

The final temperature of the mixture is 36.8°C

7. A block of iron of mass 150g and having a thermal capacity of 69J K^{-1} is observed to cool at a rate of 0.15°C per second when placed in a refrigerator. Calculate:

a. the rate at which the block is losing heat
b. the specific heat capacity of the iron.

Solutions

a. The rate at which the block is losing heat is given by: $\frac{Q}{t}$

But $Q = mc\Delta\theta$ and $mc = C$ (Where C is the heat/thermal capacity)

$\therefore Q = C\Delta\theta$ (When C is substituted for mc)

$$\frac{Q}{t} = \frac{C\Delta\theta}{t}$$

$= 69 \times \frac{0.15}{1 \text{ sec}}$ (Note that 0.15°C is the change in temperature, (i.e. $\Delta\theta$) in 1 second)

$\frac{Q}{t} = 10.35 \text{J/sec}$

The block is losing heat at a rate of 10.35J/sec

b. $C = mc$

$69 = (\frac{150}{1000}) \times c$ (The mass, m, must be expressed in Kg by diving by 1000)

$69 = 0.15c$

$\therefore c = \frac{69}{0.15} = 460$

The specific heat capacity of the iron is 460JKg^{-1}K^{-1}

8. How long will it take a 750W heater operating at full rating to raise the temperature of 1.5Kg of water from 40°C to 70°C? (Specific heat capacity of water = 4200J/KgK)

Solution

$Q = IVt = mc\Delta\theta$

$\therefore IVt = mc\Delta\theta$

Substituting P for IV gives:

$$Pt = mc\Delta\theta$$

$$750t = 1.5 \times 4200 \times (70 - 40)$$

$$750t = 1.5 \times 4200 \times 30$$

$$750t = 189000$$

$$\therefore t = \frac{189000}{750}$$

$$t = 252 \text{ sec}$$

It will take 252 seconds.

9. A current of 4A at a voltage supply of 220V is used to supply heat to a certain quantity of oil at a temperature of 24°C. If the oil was heated for 2 minutes to attain a temperature of 49°C, what is the mass of the oil? (Specific heat capacity of oil = $2170 \text{JKg}^{-1}\text{K}^{-1}$)

Solution

2 minutes = 2 x 60 = 120sec   (Since 60 seconds = 1 Minute)

$$IVt = mc\Delta\theta$$

$$4 \times 220 \times 120 = m \times 2170 \times (49 - 24)$$

$$105600 = m \times 2170 \times 25$$

$$105600 = 54250m$$

$$\therefore m = \frac{189000}{750}$$

$$m = 1.95 \text{Kg}$$

The mass of the oil is 1.95Kg

10. An electric heater rated 500W is used to heat a copper block of 250g in 20sec. Calculate the rise in temperature of the copper. (Specific heat capacity of copper is 380JKg^{-1}K^{-1})

Solution

Since the unit of mass in 250g is g and the unit of mass in 380JKg^{-1}K^{-1} is Kg, then we convert 250g to Kg to make the units the same.

∴ 250g = ($\frac{250}{1000}$)Kg = 0.25Kg

∴ Pt = mcΔθ

500 x 20 = 0.25 x 380 x Δθ

10000 = 95Δθ

Δθ = $\frac{10000}{95}$

= 105.3

The rise in temperature of the copper is 105.3°C

Exercise 16

1. How much heat is absorbed when a piece of iron of mass 200g and specific heat capacity 450JKg^{-1}K^{-1}, is heated from 25°C to 90°C?

2. Hot water at a temperature of T°C is added to thrice that amount of water at a temperature at 25°C. If the resulting temperature of the mixture is 42°C, calculate the value of T.

3. Water of mass 1200g at a temperature of 82°C is added to 480g of water at 26°C. What is the final temperature of the mixture?

4. A man wishes to bathe with warm water at 40°C. What mass of hot water at 85°C should he add to 18kg of cold water at 21°C in order to obtain the warm.

5. A piece of copper of mass 50g at 240°C is transferred into a copper calorimeter of mass 40g containing 70 g of water at 28°C. Neglecting heat loses, what will be the final temperature of the mixture? (Specific heat capacity of copper and water are $0.4 Jg^{-1}K^{-1}$ and $4.2 Jg^{-1}K^{-1}$ respectively)

6. A piece of copper of mass 32g at 220°C is placed in a copper calorimeter of mass 45g containing 52 g of water at 24°C. Neglecting heat loses to the surroundings, calculate the final temperature of the mixture? (Specific heat capacity of water = $4200 JKg^{-1}K^{-1}$, Specific heat capacity of copper = $400 JKg^{-1}K^{-1}$)

7. A block of iron of mass 90g and having a thermal capacity of 65J K^{-1} is observed to cool at a rate of 0.2°C per second when placed in a refrigerator. Calculate:

a. the rate at which the block is losing heat
b. the specific heat capacity of the iron.

8. How long will it take a 1000W heater operating at full rating to raise the temperature of 3.2Kg of water from 35°C to 85°C? (Specific heat capacity of water = 4200J/KgK)

9. A current of 6.5A at a voltage supply of 240V is used to supply heat to a certain quantity of oil at a temperature of 30°C. If the oil was heated for 2.2 minutes to attain a temperature of 64°C, what is the mass of the oil? (Specific heat capacity of oil = $2200 JKg^{-1}K^{-1}$)

10. An electric heater rated 750W is used to heat a copper block of 500g in 45sec. Calculate the rise in temperature of the copper. (Specific heat capacity of copper is $400 JKg^{-1}K^{-1}$)

11. How long will it take a 1.2KW heater operating at full rating to raise the temperature of 240g of water from 18°C to 100°C? (Specific heat capacity of water = 4200J/KgK)

12. Hot water at a certain temperature is added to four times that amount of water at a temperature at 29°C. If the resulting temperature of the mixture is 38°C, calculate the temperature of the hot water.

CHAPTER 17
CHANGE OF STATE – LATENT HEAT AND SPECIFIC LATENT HEAT

During the change of state of a substance, there is no change in temperature of the substance. The quantity of heat energy needed for this change of state is call latent heat. It is given by:

$$Q = mL$$

Where Q is the quantity of heat (i.e. the latent heat), m is the mass of the substance, while L is the specific latent heat of the substance.

Note that when there is a change in temperature of a substance when it is heated or cooled, the quantity of heat energy remains $Q = mc\Delta\theta$. It is only at change of state that $Q = mL$.

Examples

1. Determine the heat required to melt 30g of ice at 0°C. (Specific latent heat of fusion of ice = 335J/g)

Solution

Since the unit of mass is g in 30g and 335J/g then there is no need of converting the mass from g to Kg.

The heat required to melt the ice is given by:
$$Q = mL$$
$$= 30 \times 335$$
$$= 10500J$$

This can be divided by 1000 to convert Joules to Kilojoules. This gives:
$$= 10.5KJ$$

The heat required to melt the ice is 10.5KJ.

2. Determine the heat required to change 10g of ice at 0°C to water at 20°C. (Specific latent heat of fusion of ice = 330J/g, Specific heat capacity of water = 4.19Jg^{-1}K^{-1})

Solution

Here, the total heat is the heat required to melt the ice + the heat required to heat the melted ice (now water) from 0°C to 20°C.

∴ Q = mL + mcΔθ

= (10 x 330) + [10 x 4.19 x (20 – 0)]

Note that when the 10g ice melts, the mass of water formed remains 10g

= 3300 + (41.9 x 20)

= 3300 + 838

= 4138

The heat required is 4138J

3. Calculate the mass of ice at its melting point that would be just melted by 500g of boiling water. (Specific heat capacity of water = 4.2Jg^{-1}K^{-1}, Specific latent heat of fusion of ice = 336J/g)

Solution

This is a case of a mixture of hot and cold body. The heat lost by the boiling water in cooling from 100°C to 0°C is equal to the heat gained by the ice in melting it.

∴ (mL)$_{ice}$ = (mcΔθ)$_{boiling\ water}$

m x 336 = 500 x 4.2 x (100 – 0)

336m = 2100 x 100

$m = \dfrac{210000}{336}$

m = 625g

The mass of ice that would melt is 625g

4. A copper calorimeter of mass 0.11Kg and specific heat capacity 400JKg^{-1}K^{-1} contains 94g of water at 35°C. When 10g of ice at 0°C is added and allowed to melt, the temperature falls to 25°C. Calculate the value of the specific latent heat of fusion of ice. (Specific heat capacity of water = 4200JKg^{-1}K^{-1}).

Solutions

Convert the masses in g to Kg, as follows:

$$94g = \frac{90}{1000} = 0.094Kg$$

$$10g = \frac{10}{1000} = 0.01Kg$$

∴ Heat gained by the ice to melt + Heat gained by the water formed from the melted ice to rise in temperature from 0°C to 25°C = Heat lost by water at 35°C + Heat lost by calorimeter.

This gives the expression:

$(mL)_{ice}$ + $(mc\Delta\theta)_{water\ from\ ice}$ = $(mc\Delta\theta)_w$ + $(mc\Delta\theta)_c$ (Where w represents water and c represents calorimeter)

Substituting the appropriate values into the expression above gives:

(0.01 x L) + 0.01 x 4200 x (25 – 0) = [0.094 x 4200 x (35 – 25)] + [0.11 x 400 x (35 – 25)]

0.01L + (42 x 25) = (394.8 x 10) + (44 x 10)

0.01L + 1050 = 3948 + 440

0.01L = 4388 – 1050

0.01L = 3338

∴ $L = \frac{3338}{0.01}$

L = 333800

The specific latent heat of fusion of ice is 333800J/Kg

5. Calculate the heat required to change 0.02Kg of ice at -10°C to steam at 100°C. (Specific heat capacity of ice = 2.14Jg^{-1}K^{-1}, specific latent heat of fusion of ice = 330J/g, specific heat capacity of water = 4.2Jg^{-1}K^{-1}, specific latent heat of vaporization of steam = 2260J/g)

Solution

The mass of ice is given in Kg, while the masses in the constants given are in g. So we convert the 0.02Kg to g to make the units the same. This gives:

0.02Kg = 0.02 x 1000 = 20g

The total heat required = Heat needed to heat the ice from -10°C to 0°C + heat needed to melt the ice at 0°C + heat needed to heat the water formed from 0°C to 100°C + heat needed to vaporize the boiling water at 100°C.

Note that there are two change of state (i.e. mL) and two change in temperature (i.e. mcΔθ).

The total heat is thus given by the expression below.

Q = (mcΔθ)$_{ice}$ + (mL)$_{ice}$ + (mcΔθ)$_w$ + (mL)$_{steam}$

= 20 x 2.14 x [0 −(-10)] + (20 x 330) + [20 x 4.2 x (100 − 0)] + (20 x 2260)

= (20 x 2.14 x 10) + 660 + (20 x 4.2 x 100) + 45200

= 428 + 660 + 8400 + 45200

= 60628

The total heat required is 60628J or 60.628KJ

6. What amount of current would pass through a 10Ω coil if it takes 21sec for the coil to just melt a lump of ice of mass 10g at 0°C. (Specific latent heat of fusion of ice = 336J/g)

Solution

The electrical energy supplied = The latent heat needed to melt the ice

∴ $I^2Rt = mL$

$I^2 \times 10 \times 21 = 10 \times 336$

$210I^2 = 3360$

$I^2 = \dfrac{3360}{210}$

$I^2 = 16$

$I = \sqrt{16}$

$I = 4$

The amount of current is 4A

7. What is the voltage supply when a 7Ω coil heats 20g of ice to water at 5°C in 25sec? (L_{ice} = 336J/g, specific heat capacity of water = 4.2Jg^{-1}K^{-1})

Solution

The total amount of heat needed to melt the ice and heat the water formed from 0°C to 5°C is given by: $mL + mc\Delta\theta$

This amount of heat will be equal to the electrical energy provided (i.e. V^2t/R)

∴ $mL + mc\Delta\theta = V^2t/R$

$(20 \times 336) + [20 \times 4.2 \times (5 - 0)] = V^2 \times \dfrac{25}{7}$

$6720 + 420 = \frac{25}{7}V^2$

$7140 = \frac{25}{7}V^2$

$7 \times 7140 = 25V^2$

$\therefore V^2 = \frac{7 \times 7140}{25}$

$= 1999.2$

$V = \sqrt{1999.2}$

$V = 44.7$

The voltage supply is 44.7v

Exercise 17

1. Determine the heat required to melt 50g of ice at 0°C. (Specific latent heat of fusion of ice = 336J/g)

2. Determine the heat required to change 15g of ice at 0°C to water at 25°C. (Specific latent heat of fusion of ice = 330J/g, Specific heat capacity of water = $4.18Jg^{-1}K^{-1}$)

3. Calculate the mass of ice at its melting point that would be just melted by 250g of boiling water. (Specific heat capacity of water = $4.2Jg^{-1}K^{-1}$, Specific latent heat of fusion of ice = 335J/g)

4. A copper calorimeter of mass 0.15Kg and specific heat capacity $400JKg^{-1}K^{-1}$ contains 88g of water at 27°C. When 12g of ice at 0°C is added and allowed to melt, the temperature falls to 18°C. Calculate the value of the specific latent heat of fusion of ice. (Specific heat capacity of water = $4200JKg^{-1}K^{-1}$).

5. Calculate the heat required to change 0.19Kg of ice at -5°C to steam at 100°C. (Specific heat capacity of ice = $2.15Jg^{-1}K^{-1}$, specific latent heat of fusion of ice = 330J/g,

specific heat capacity of water = 4.2Jg^{-1}K^{-1}, specific latent heat of vaporization of steam = 2260J/g)

6. What amount of current would pass through a 20Ω coil if it takes 30sec for the coil to just melt a lump of ice of mass 60g at 0°C. (Specific latent heat of fusion of ice = 330J/g)

7. What is the voltage supply when a 10Ω coil heats 50g of ice at 0°C to water at 12°C in 2 minutes? (L$_{ice}$ = 335J/g, specific heat capacity of water = 4.2Jg^{-1}K^{-1})

8. A copper calorimeter of mass 150g and specific heat capacity 400JKg^{-1}K^{-1} contains 14g of ice at 0°C. What will be the final temperature of the mixture when 500g of water at 80°C is added to the calorimeter? (Specific heat capacity of water = 4200JKg^{-1}K^{-1}, Specific latent heat of fusion of ice = 330J/g).

9. A piece of copper of mass 450g at 180°C is placed in a copper calorimeter of mass 65g containing 20g of ice at 0°C. Neglecting heat loses to the surroundings, calculate the final temperature of the mixture? (Specific heat capacity of water = 4200JKg^{-1}K^{-1}, Specific heat capacity of copper = 400 JKg^{-1}K^{-1}, Specific latent heat of fusion of ice = 330000J/Kg)

10. How long will it take a coil of 40Ω to just melt a lump of ice of mass 30g at 0°C. if the voltage supply is 220V. (Specific latent heat of fusion of ice = 330J/g)

CHAPTER 18
RELATIVE HUMIDITY

The amount of water vapour present in the atmosphere is referred to as humidity. Relative humidity is a term used to describe how humid the air is. Thus, relative humidity is defined as a ratio as follows:

$$R.H = \frac{\text{Mass of water vapour present in a certain volume of air at a temperature}}{\text{Mass of water vapour needed to saturate the volume of air at that temperature}}$$

Where R.H represents relative humidity. This fraction is multiplied by 100 to give the relative humidity as percentage.

DEW POINT

The temperature at which the water vapour present in the air is just sufficient to saturate it, is called dew point.

Relative humidity can also be defined in terms of dew point as expressed below:

$$R.H = \frac{\text{Saturated vapour pressure of water at the dew point}}{\text{Saturated vapour pressure of water at the air temperature}}$$

Examples

1. The mass of water vapour needed to saturate 25cm³ of air at 28°C is 4g. Calculate the relative humidity of the air if the mass of water vapour present in the air is 3g.

Solution

$$R.H = \frac{\text{Mass of water vapour in the air}}{\text{Mass of water vapour needed to saturate the air}} \times 100$$

$$= \frac{3}{4} \times 100$$

= 75

The relative humidity of the air is 75%

2. The mass of water vapour needed to saturate a certain volume of air at a particular temperature is 5.5g. If the relative humidity of the air is 60%, calculate the mass of water vapour present in the air.

Solution

$$R.H = \frac{\text{Mass of water vapour in the air}}{\text{Mass of water vapour needed to saturate the air}} \times 100$$

Let the mass of water vapour present in the air be m

$$\therefore 60 = \frac{m}{5.5} \times 100$$

$$60 = \frac{100m}{5.5}$$

$$60 \times 5.5 = 100m$$

$$\therefore m = \frac{60 \times 5.5}{100}$$

= 3.3

The mass of water vapour present in the air is 3.3g.

3. The dew point of air is found to be 15°C when the air temperature is 25°C. Calculate the relative humidity if the saturated vapour pressure of water at 15°C and 25°C are 1.25cm and 2.34cm of mercury respectively.

Solution

$$R.H = \frac{\text{Saturated vapour pressure of water at the dew point}}{\text{Saturated vapour pressure of water at the air temperature}} \times 100$$

$$= \frac{1.25}{2.34} \times 100$$

$$= 53.4$$

The relative humidity is 53.4%

4. The saturated vapour pressure of water at 22°C is 2.08cmHg when the air temperature is 36°C. If the relative humidity of the air is 82%, calculate the saturated vapour pressure of water at 36°C given that the dew point of air is 22°C.

Solution

$$R.H = \frac{\text{Saturated vapour pressure of water at the dew point}}{\text{Saturated vapour pressure of water at the air temperature}} \times 100$$

Let the saturated vapour pressure of water be x

$$82 = \frac{2.08}{x} \times 100$$

$$82 = \frac{2.08 \times 100}{x}$$

$$82x = 2.08 \times 100$$

$$\therefore x = \frac{2.08 \times 100}{82}$$

$$x = 2.54$$

The saturated vapour pressure of water is 2.54cmHg

5. The mass of water vapour needed to saturate a certain volume of air at a particular temperature is 7.6g. If the relative humidity of the air is 0.56, calculate the mass of water vapour present in the air.

Solution

$$R.H = \frac{\text{Mass of water vapour in the air}}{\text{Mass of water vapour needed to saturate the air}} \times 100$$

Since the relative humidity is given in decimal and not in percentage, it means that it was not multiplied by 100. So the formula will not be multiplied by 100 as follows:

$$R.H = \frac{\text{Mass of water vapour in the air}}{\text{Mass of water vapour needed to saturate the air}}$$

Let the mass of water vapour present in the air be m

$$\therefore 0.56 = \frac{m}{7.6}$$

$$m = 0.56 \times 7.6$$

$$= 4.256$$

The mass of water vapour present in the air is 4.256g.

6. The saturated vapour pressure of water at 18°C is 1.94cmHg when the air temperature is 29°C. If the relative humidity of the air is 0.45, calculate the saturated vapour pressure of water at 29°C given that the dew point of air is 18°C.

Solution

Since the relative humidity is given in decimal and not in percentage, then it expressed as:

$$R.H = \frac{\text{Saturated vapour pressure of water at the dew point}}{\text{Saturated vapour pressure of water at the air temperature}}$$

Let the saturated vapour pressure of water be x

$$0.45 = \frac{1.94}{x}$$

$$0.45x = 1.94$$

$$\therefore \quad x = \frac{1.94}{0.45}$$

$$x = 4.31$$

The saturated vapour pressure of water is 4.31cmHg

Exercise 18

1. The mass of water vapour needed to saturate 20cm^3 of air at 25°C is 6g. Calculate the relative humidity of the air if the mass of water vapour present in the air is 4g.

2. The mass of water vapour needed to saturate a certain volume of air at a particular temperature is 8.2g. If the relative humidity of the air is 75%, calculate the mass of water vapour present in the air.

3. The dew point of air is found to be 18°C when the air temperature is 27°C. Calculate the relative humidity if the saturated vapour pressure of water at 18°C and 27°C are 2cm and 3.9cm of mercury respectively.

4. The saturated vapour pressure of water at 24°C is 1.95cmHg when the air temperature is 32°C. If the relative humidity of the air is 68%, calculate the saturated vapour pressure of water at 32°C given that the dew point of air is 24°C.

5. The mass of water vapour needed to saturate a certain volume of air at a particular temperature is 5.3g. If the relative humidity of the air is 0.41, calculate the mass of water vapour present in the air.

6. The saturated vapour pressure of water at 20°C is 2.01cmHg when the air temperature is 30°C. If the relative humidity of the air is 0.83, calculate the saturated vapour pressure of water at 30°C given that the dew point of air is 20°C.

CHAPTER 19
BOYLE'S LAW

The relationship between volume and pressure according to Boyle's law is:

$$V \propto 1/P$$

Or, $V = K/P$

Where V=volume of the given gas, P=pressure of the gas and K=constant.

If the volume of or pressure of a gas changes from an initial value to a final value, then,

$$P_1V_1 = P_2V_2$$

Where 1 represent the initial values, and 2 represent the final values.

The temperature of the gas must be kept constant for the law to be valid.

Practical applications of Boyle's law

The use of mercury in capillary tubes (or other kinds of tubes) can be used to determine the atmospheric pressure of a place. A length of mercury used to trapped volume of air in a tube can be used for this purpose by applying Boyle's law. The length of air column trapped in the tube is proportional to the volume of the air. It is taken to be equal to the volume of the air trapped in the tube.

The various position of the capillary tube and the values of their pressures and volumes are as given below.

When the tube is vertical with the open end upwards

In this case,

$$P = H + h$$

and   V = l

where P=pressure, V=volume, H=atmospheric pressure, h=the length of the mercury thread and l=length of the trapped air column.

When the tube is vertical with the open end downwards

In this case,

    P = H - h

and   V = l

When the tube is horizontal

In this case,

    P = H

and   V = l

Other apparatus where mercury can be used to trap a column of air include a "J" tube, an inverted tube in a trough of mercury, etc. For apparatus such as these, P = H + h, when the mercury in the open end is higher than the mercury in the closed end, and P = H - h when the mercury in the closed end is higher than the mercury in the open end. In this case, h = the difference in height between the level of the mercury in the open end and the level of the mercury in the closed end, i.e. the difference in height between the two mercury levels.

Examples

1. A thread of mercury of length 15cm is used to trap some air in a capillary tube with uniform cross sectional area and closed at one end. With the tube vertical and the

open end upwards, the length of trapped air column is 20cm. Calculate the length of the air column when the tube is held:

a. horizontally

b. vertically with the open end downwards.

   (Atmospheric pressure = 76cmHg)

Solutions

a. H = 76 and h = 15 (the length of the mercury thread)

   By applying Boyle's law, the initial condition of the air is:

   $P_1 = H + h$ (when the tube is vertical with the open end upwards)

   and $V_1 = l$

   $\therefore P_1 = 76 + 15 = 91$

   $P_1 = 91$, and $V_1 = 20$

   When the tube becomes horizontal, then the final condition of the air is:

   $P_2 = H$ (when the tube is horizontal)

   and $V_2 = ?$

   $\therefore P_2 = 76$ and $V_2 = ?$ (V_2 is to be calculated)

   Therefore, from Boyle's law,

   $P_1V_1 = P_2V_2$

   91 x 20 = 76 x V_2

   1820 = 76V_2

   $\therefore V_2 = \dfrac{1820}{76}$

= 23.95

The volume of the trapped air is 23.95cm^3

b. As before, the initial condition remains:

$P_1 = 91$, and $V_1 = 20$

When the tube becomes vertical with the open end downwards, then the final condition of the air is:

$P_2 = H - h$ (when the tube is vertical with the open end downward)

and $V_2 = ?$

$P_2 = 76 - 15 = 61$

∴ $P_2 = 61$ and $V_2 = ?$ (V_2 is to be calculated)

Therefore, from Boyle's law,

$P_1V_1 = P_2V_2$

91 x 20 = 61 x V_2

1820 = 61V_2

∴ $V_2 = \dfrac{1820}{61}$

= 29.8

The volume of the trapped air is 29.8cm^3

2. A glass tube in the form of letter J has the shorter limb sealed and the longer limb open. Mercury is poured into the tube until the level in either limb is the same when the tube is vertical. In this position the length of the air column in the sealed limb is 63cm. More mercury is then poured into the tube until the length of the trapped air

column is 42cm. Calculate the difference in the levels of mercury in the limbs if a nearby barometer reads 75cm and the temperature of the surroundings remain the same.

Solution

a.   H = 75   (the reading of the barometer gives the atmospheric pressure)
When the two mercury levels are the same, the difference in their height is = 0. Hence, the initial condition of the trapped air is:

$P_1 = H + h$

$\quad\quad = H + 0$

$P_1 = H$

and $V_1 = l$

∴ $P_1 = 75$

$V_1 = 63$

When mercury is poured into the tube, then the final condition of the air is:

$P_2 = H + h$ (where h is the difference in height between the two mercury levels)

$V_2 = l$

∴ $P_2 = 75 + h$

$V_2 = 42$

Therefore, from Boyle's law,

$P_1V_1 = P_2V_2$

75 x 63 = (75 + h) x 42

4725 = 3150 + 42h

4725 − 3150 = 42h

$$1575 = 42h$$

$$\therefore h = \frac{1575}{42}$$

$$= 37.5$$

The difference in the levels of mercury in the limbs is 37.5cm

Note that H + h was used for P_2 because the level of mercury in the open end is higher than that in the closed end. If the level in the closed end is higher than that in the open end then P_2 will be given by H − h.

3. A uniform capillary tube closed at one end contained dry air trapped by a thread of mercury 8.5cm long. When the tube was held horizontally, the length of the air column was 5cm, when it was held vertically with the closed end downward, the length was 4.5cm. Determine:

a. the value of the atmospheric pressure
b. the length of the air column when the tube is held vertically, with the open end downward.

Solution

a. H = ?

h = 8.5 (the length of the mercury thread)

The initial condition of the air is:

$P_1 = H$ (when the tube is horizontal)

$V_1 = 5$

When the tube becomes vertical with the closed end downward, then the final condition of the air is:

$P_2 = H + h$

∴ $P_2 = H + 8.5$

   $V_2 = 4.5$

Therefore, from Boyle's law,

   $P_1V_1 = P_2V_2$

   $H \times 5 = (H + 8.5) \times 4.5$

   $5H = (H + 8.5)4.5$

   $5H = 4.5H + 38.25$

   $5H - 4.5H = 38.25$

   $0.5H = 38.25$

∴ $H = \dfrac{38.25}{0.5}$

   $= 76.5$

The atmospheric pressure is 76.5cmHg

b. As before the initial condition of the air is:

   $P_1 = H$ (when the tube is horizontal)

   $P_1 = 76.5$ (As obtained from question (a) above)

   $V_1 = 5$

When the tube becomes vertical with the open end downward, then the final condition of the air is:

   $P_2 = H - h$

   $= 76.5 - 8.5$

∴ $P_2 = 68$

$V_2 = ?$

Therefore, from Boyle's law,

$P_1V_1 = P_2V_2$

$76.5 \times 5 = 68 \times V_2$

$382.5 = 68V_2$

$\therefore V_2 = \dfrac{382.5}{68}$

$= 5.6$

The volume of the trapped air is $5.6 cm^3$

4. A fixed mass of a gas occupies a volume of $50 cm^3$ at 710mmHg. Calculate its volume when the pressure increases to 800mmHg at constant temperature.

Solutions

Given from the question are: $P_1 = 710$, $V_1 = 50$, $P_2 = 800$, $V_2 = ?$
Since the temperature was kept constant, then it means that Boyle's will be applied.

$\therefore P_1V_1 = P_2V_2$

$710 \times 50 = 800 \times V_2$

$35500 = 800V_2$

$\therefore V_2 = \dfrac{35500}{800}$

$V_2 = 44.4 cm^3$

The volume of the gas is $44.4 cm^3$

5. A fixed mass of a gas occupies a volume of 120cm³ at 72cmHg. Calculate its pressure when the volume becomes 190cm³.

Solution

Given from the question are: P_1 = 72, V_1 = 120, V_2 = 190, P_2 = ? Since nothing was said about the temperature, then it is assumed that the temperature was kept constant. So, from Boyle's:

$P_1V_1 = P_2V_2$

72 x 120 = P_2 x 190

72 x 120 = 190P_2

∴ $P_2 = \dfrac{72 \times 120}{190}$

P_2 = 45.5

The pressure of the gas is 45.5cmHg

6. At a temperature of 70°C, a given mass of gas at 650mmHg was expanded to four times its original volume at the same temperature. Calculate its new pressure.

Solution

At the same temperature means that the temperature, 70°C was kept constant. So, this is Boyle's law (Constant temperature).

Given from the question are: P_1 = 650, Initial volume = V_1 (Since the value was not given), V_2 = $4V_1$ (Expanded to four times its original volume), P_2 = ? Note that the constant value of the temperature will not be used, since it did not change.

So, from Boyle's law:

$P_1V_1 = P_2V_2$

$650 \times V_1 = P_2 \times 4V_1$

$650V_1 = 4V_1P_2$

Cancelling out the V_1 when both sides are divided by V_1 gives:

$650 = 4P_2$

$\therefore P_2 = \dfrac{650}{4}$

$P_2 = 162.5$

The new pressure of the gas is 162.5cmHg

7. A fixed mass of gas at a volume of 440cm^3 was expanded such that its original pressure was halved. Calculate its new volume if the temperature is constant.

Solution

Given from the question are: Initial pressure = P_1 (Since the value was not given), V_1 = 440, $P_2 = \frac{1}{2}P_1$ (Since the original pressure was halved), V_2=?
So, from Boyle's:

$P_1V_1 = P_2V_2$

$P_1 \times 440 = \frac{1}{2}P_1 \times V_2$

$440P_1 = P_1V_2/2$

Cancelling out the P_1 when both sides are divided by P_1 gives:

$440 = V_2/2$

$\therefore V_2 = 440 \times 2$

$V_2 = 880$

The new volume of the gas is 880cm^3

It is obvious from examples 6 and 7 that there is a direct method of solving Boyle's law when one quantity is increased or decreased by a number of times. Therefore, we can conclude from example 6 that when volume is increased four times, then pressure is reduced four times. This also means that when volume is increased by say two times, then pressure is reduced by two times. We can conclude from example 7 that when pressure is reduced by two times (i.e. when it is halved), then the volume is increased by two times. This also means that when pressure is reduced by say five times, then the volume is increased by five times.

Generally, when one quantity is increased by a number of times, the other quantity is reduced by the same number of times. Or, when one quantity is reduced by a number of times, then the other quantity is increased by the same number of times. This is based on the fact that in Boyle's law the pressure and volume are inversely related.

8. Calculate the new volume of a fixed mass of gas at 140cm^3 when its pressure is increased seven times at constant temperature.

Solution

When the pressure is increased seven times, it means that the volume will be reduced seven times.

∴ The new volume is = $\frac{140}{7}$

= 20cm^3

9. What is the new pressure of a fixed mass of gas at 330mmHg when its volume is decreased three times at constant temperature?

Solution

When the volume is decreased three times, it means that the pressure will be increased three times.

∴ The new pressure is = 330 x 3

= 990mmHg

10. A given mass of gas occupies a volume of 500cm³ at a particular temperature. How will the pressure be affected if the gas is compressed to 25cm³ at the same temperature?

Solution

When the volume is reduced from 500cm³ to 25cm³, it means that it has been reduced by a factor that is given by:

$$\frac{500}{25} = 20$$

This means that the volume has been reduced twenty times.

Therefore the pressure will be increased twenty times.

Exercise 19

1. A thread of mercury of length 20cm is used to trap some air in a capillary tube with uniform cross sectional area and closed at one end. With the tube vertical and the open end upwards, the length of trapped air column is 18cm. Calculate the length of the air column when the tube is held:
a. horizontally

b. vertically with the closed end upwards.

(Atmospheric pressure = 76cmHg)

2. A glass tube in the form of letter J has the shorter limb sealed and the longer limb open. Mercury is poured into the tube until the level in either limb is the same when the tube is vertical. In this position the length of the air column in the sealed limb is 58cm. More mercury is then poured into the tube until the length of the trapped air

column is 30cm. Calculate the difference in the levels of mercury in the limbs if a nearby barometer reads 76cm and the temperature of the surroundings remains constant.

3. A uniform capillary tube closed at one end contained dry air trapped by a thread of mercury 10cm long. When the tube was held horizontally, the length of the air column was 6.5cm. When it was held vertically with the closed end downward, the length was 5.8cm. Determine:
a. the value of the atmospheric pressure
b. the length of the air column when the tube is held vertically, with the open end downward.

4. A fixed mass of a gas occupies a volume of 400cm^3 at 690mmHg. Calculate its volume when the pressure increases to 880mmHg at constant temperature.

5. A fixed mass of a gas occupies a volume of 200cm^3 at 77cmHg. Calculate its pressure when the volume becomes 90cm^3.

6. At a temperature of 30°C, a given mass of gas at 600mmHg was expanded to three times its original volume at the same temperature. Calculate its new pressure.

7. A fixed mass of gas at a volume of 1700mm^3 was expanded such that its original pressure was quartered. Calculate its new volume if the temperature is constant.

8. Calculate the new volume of a fixed mass of gas at 620cm^3 when its pressure is increased five times at constant temperature.

9. What is the new pressure of a fixed mass of gas at 900mmHg when its volume is doubled at constant temperature?

10. A given mass of gas occupies a volume of 200cm^3 at a particular temperature. How will the pressure be affected if the gas is compressed to 10cm^3 at the same temperature?

11. At a temperature of 120°C, a given mass of gas at 810mmHg was expanded to five times its original volume at the same temperature. Calculate its new pressure.

12. A fixed mass of a gas occupies a volume of 280cm^3 at 740mmHg. Calculate its volume when the pressure increases to 1200mmHg at constant temperature.

CHAPTER 20
CHARLES' LAW

The relationship between volume and temperature according to Charles' law is:

$V \alpha T$

Or, $V = KT$

Where V=volume of the given gas, T= temperature of the gas in Kelvin, and K=constant.

If the volume or temperature of a gas changes from an initial value to a final value, then,

$$\frac{V_1}{T_1} = \frac{V_2}{T_2}$$

Where 1 represent the initial values, and 2 represent the final values.

The pressure of the gas must be kept constant for the law to be valid.

Note that in order to convert temperature from °C to Kelvin, simply add 273. Similarly, to convert from K to °C, subtract 273.

Examples

1. A fixed mass of gas occupies a volume of 240cm³ at 27°C. Calculate its volume when the temperature is increased to 42°C at constant pressure.

Solution

The given values are: V_1 = 240, T_1 = 27 + 273 = 300, T_2 = 42 + 273 = 315, V_2 = ? (The temperatures in °C have been converted to temperatures in K by adding 273)
∴ V_1 = 240, T_1 = 300, T_2 = 315, V_2 = ?

By applying Charles' law:

$$\frac{V_1}{T_1} = \frac{V_2}{T_2}$$

$$\frac{240}{300} = \frac{V_2}{315}$$

Cross multiply to get:

$300V_2 = 240 \times 315$

$$V_2 = \frac{240 \times 315}{300}$$

$= 252$

The volume is 252cm^3

2. A fixed mass of a gas occupies a volume of 90cm^3 at 72^{0}C. Calculate its temperature when the volume increases to 250cm^3.

Solution

Given from the question are: V_1 = 90, T_1 = 72 + 273 = 345, V_2 = 250, T_2 = ? Since nothing was said about the pressure, then it is assumed that the pressure was kept constant. So, from Charles' law:

$$\frac{V_1}{T_1} = \frac{V_2}{T_2}$$

$$\frac{90}{345} = \frac{250}{T_2}$$

Cross multiply to get:

$90T_2 = 345 \times 250$

$$T_2 = \frac{345 \times 250}{90}$$

$= 958$

The Temperature is 958K

3. Air at a temperature of 47^{0}C occupies a volume of 380cm^3. Calculate the temperature of the air in ^{0}C when its volume becomes 150cm^3 at constant pressure.

Solution

Given from the question are: $T_1 = 47 + 273 = 320$, $V_1 = 380$, $V_2 = 150$, $T_2 = ?$.
By applying Charles' law:

$$\frac{V_1}{T_1} = \frac{V_2}{T_2}$$

$$\frac{380}{320} = \frac{150}{T_2}$$

$380 T_2 = 320 \times 150$

$$T_2 = \frac{320 \times 150}{380}$$

$\qquad = 126K$ (This is temperature in Kelvin)

$\therefore$ Temperature in $^0C = 126 - 273 = -147$

The temperature of the air is -147^0C

4. A certain gas at a temperature of -120^0C occupies a volume of $220cm^3$. Calculate the temperature of the gas in ^0C when its volume is increased to $740cm^3$ at the same pressure.

Solution

Given from the question are: $V_1 = 220$, $T_1 = -120 + 273 = 153$, $V_2 = 740$, $T_2 = ?$.
By applying Charles' law:

$$\frac{V_1}{T_1} = \frac{V_2}{T_2}$$

$$\frac{220}{153} = \frac{740}{T_2}$$

$220 T_2 = 153 \times 740$

$$T_2 = \frac{153 \times 740}{220}$$

= 515K (This is temperature in Kelvin)

∴ Temperature in °C = 515 – 273 = 242

The temperature of the gas is 242°C

5. A fixed mass of gas occupies a volume of 165cm³ at 324K. Calculate its volume when the temperature is increased to 400K at constant pressure.

Solution

The given values are: V_1 = 165, T_1 = 324, T_2 = 400, V_2 = ? (The temperatures are already given in K, so there is no need for any conversion)

By applying Charles' law:

$$\frac{V_1}{T_1} = \frac{V_2}{T_2}$$

$$\frac{165}{324} = \frac{V_2}{400}$$

$324 V_2 = 165 \times 400$

$$V_2 = \frac{165 \times 400}{324}$$

= 204

The volume of the gas is 204cm³

6. A certain volume of air has a temperature of 240K at a pressure of 1200mmHg. Calculate the temperature of the air when the volume is tripled at 1200mmHg.

Solution

The pressure of 1200mmHg did not change (i.e. it was kept constant). So, this represents Charles' law.

The given conditions are: Initial volume = V_1 (Since the value is not given), T_1 = 240, V_2 = $3V_1$ (Since the initial volume is tripled to give final volume), T_2 = ?

From Charles' law:

$$\frac{V_1}{T_1} = \frac{V_2}{T_2}$$

$$\frac{V_1}{240} = \frac{3V_1}{T_2}$$

$V_1T_2 = 240 \times 3V_1$

Dividing both sides by V_1 will cancel out V_1 to give:

$T_2 = 240 \times 3$

$T_2 = 720$

The temperature of the air is 720K

7. A gas occupies a volume of 0.245dm³ at a given temperature. Calculate the volume of the gas when the temperature is quadrupled at constant pressure.

Solution

The given conditions are: V_1 = 0.245, Initial temperature = T_1 (Since the value is not given), $T_2 = 4T_1$ (Since the initial temperature is quadrupled, i.e. multiplied by four to give the final temperature), V_2 = ?

From Charles' law:

$$\frac{V_1}{T_1} = \frac{V_2}{T_2}$$

$$\frac{0.245}{T_1} = \frac{V_2}{4T_1}$$

Cross multiply to get:

$V_2T_1 = 0.245 \times 4T_1$

Dividing both sides by T_1 will cancel out T_1 to give:

$V_2 = 0.245 \times 4$

$V_2 = 0.98$

The volume of the gas is $0.98 dm^3$

Examples 6 and 7 show that there is a direct method of solving Charles' law when one quantity is increased or decreased by a number of times. Therefore, we can conclude from example 6 that when volume is increased three times, then temperature is also increased three times. This also means that when volume is increased by say ten times, then temperature is also increased by ten times. We can conclude from example 7 that when temperature is increased by four times, then the volume is increased by four times also. This also means that when temperature is reduced by say seven times, then the volume is also reduced by seven times.

Generally, when one quantity is increased by a number of times, the other quantity is also increased by the same number of times. Or, when one quantity is reduced by a number of times, then the other quantity is also reduced by the same number of times. This is based on the fact that in Charles' law the volume and temperature are directly proportional to each other.

8. The volume of a fixed mass of gas is $85 cm^3$. Determine its volume when its temperature is increased eight times at constant pressure.

Solution

When the temperature is increased eight times, it means that the volume will also increase eight times.

∴ The new volume is = 85 x 8

$= 680 cm^3$

9. What is the new temperature (in °C) of a fixed mass of gas originally at 930K when its volume is decreased six times at the same pressure?

Solution

When the volume is decreased six times, it means that the temperature will also decrease six times.

∴ The new temperature is = $\dfrac{930}{6}$

= 155K

∴ The new temperature in °C = 155 – 273 = -118°C

10. The temperature of a given mass of air is 127°C. Determine its temperature in °C if its volume is doubled at the same pressure?

Solution

Temperature in Kelvin = 127 + 273 = 400K

When the volume is doubled, it means that the temperature (in Kelvin) will also be doubled.

∴ The new temperature is = 400 x 2

= 800K

∴ The new temperature in °C = 800 – 273 = 527°C

Exercise 20

1. A fixed mass of gas occupies a volume of 500cm^3 at 47°C. Calculate its volume when the temperature is increased to 87°C at constant pressure.

2. A fixed mass of a gas occupies a volume of 180cm³ at 61°C. Calculate its temperature when the volume increases to 380cm³.

3. Air at a temperature of 27°C occupies a volume of 2600mm³. Calculate the temperature of the air in °C when its volume becomes 5000mm³ at constant pressure.

4. A certain gas at a temperature of -60°C occupies a volume of 100cm³. Calculate the temperature of the gas in °C when its volume is increased to 240cm³ at the same pressure.

5. A fixed mass of gas occupies a volume of 82cm³ at 400K. Calculate its volume when the temperature is reduced to 260K at constant pressure.

6. A certain volume of air has a temperature of 350K at a pressure of 830mmHg. Calculate the temperature of the air when the volume is quadrupled at 830mmHg.

7. A gas occupies a volume of 1.05dm³ at a given temperature in Kelvin. Calculate the volume of the gas when the temperature in Kelvin is halved at constant pressure.

8. The volume of a fixed mass of gas is 125cm³. Determine its volume when its temperature in Kelvin is increased to six times its original value at constant pressure.

9. What is the new temperature (in °C) of a fixed mass of gas originally at 620K when its volume is decreased four times at the same pressure?

10. The temperature of a given mass of air is 237°C. Determine its temperature in °C if its volume is tripled at the same pressure?

CHAPTER 21
PRESSURE LAW

The relationship between pressure and temperature at constant volume is given by the pressure law as:

$P \alpha T$

Or, $P = KT$

Where P=pressure of the given gas, T= temperature of the gas in Kelvin, and K=constant.

If the pressure or temperature of a gas changes from an initial value to a final value, then,

$$\frac{P_1}{T_1} = \frac{P_2}{T_2}$$

Where 1 represent the initial values, and 2 represent the final values.

The volume of the gas must be kept constant for the law to be valid.

Note that the temperature must be expressed in Kelvin just like in Charles' law.

Examples

1. A fixed mass of gas occupies a pressure of 320mmHg at 100°C. Calculate its pressure when the temperature is increased to 152°C at constant volume.

Solution

The given values are: P_1 = 320, T_1 = 100 + 273 = 373, T_2 = 152 + 273 = 425, P_2 = ?

By applying pressure law:

$$\frac{P_1}{T_1} = \frac{P_2}{T_2}$$

$$\frac{320}{373} = \frac{P_2}{425}$$

Cross multiply to get:

373P$_2$ = 320 x 425

$$V_2 = \frac{320 \times 425}{373}$$

= 252

The pressure is 364mmHg

2. A fixed mass of a gas occupies a pressure of 54cmHg at 135^{0}C. Calculate its temperature when its pressure becomes 30cmHg.

Solution

Given from the question are: P$_1$ = 54, T$_1$ = 135 + 273 = 408, P$_2$ = 30, T$_2$ = ? The volume is not mentioned, so it is assumed to be constant.

By applying pressure law:

$$\frac{P_1}{T_1} = \frac{P_2}{T_2}$$

$$\frac{54}{408} = \frac{30}{T_2}$$

Cross multiply to get:

54T$_2$ = 408 x 30

$$T_2 = \frac{408 \times 30}{54}$$

= 227

The Temperature is 227K

3. Air at a temperature of 227°C occupies a pressure of 800mmHg. Calculate the temperature of the air in °C when its pressure is reduced to 550mmHg at constant volume.

Solution

Given from the question are: T_1 = 227 + 273 = 500, P_1 = 800, P_2 = 550, T_2 = ?.
By applying pressure law:

$$\frac{P_1}{T_1} = \frac{P_2}{T_2}$$

$$\frac{800}{500} = \frac{550}{T_2}$$

$800T_2 = 550 \times 500$

$$T_2 = \frac{550 \times 500}{800}$$

$\quad$ = 344K

∴ Temperature in °C = 344 − 273 = 71
The temperature of the air is 71°C

4. A given mass of gas at a temperature of -190°C occupies a pressure of 98cmHg. Calculate the temperature of the gas in °C when its pressure is increased to 142cmHg at the same volume.

Solution

Given from the question are: P_1 = 98, T_1 = -190 + 273 = 83, P_2 = 142, T_2 = ?.
By applying Pressure law:

$$\frac{P_1}{T_1} = \frac{P_2}{T_2}$$

$$\frac{98}{83} = \frac{142}{T_2}$$

$98T_2 = 142 \times 83$

$T_2 = \dfrac{142 \times 83}{98}$

= 120K (This is temperature in Kelvin)

∴ Temperature in °C = 120 − 273 = 153

The temperature of the gas is 153°C

5. A fixed mass of gas occupies a pressure of 2.5atm at 210K. Calculate its pressure when the temperature is increased to 360K at constant volume.

Solution

The given values are: P_1 = 2.5, T_1 = 210, T_2 = 360, P_2 = ? (The temperatures given need no conversion since they are already in Kelvin)

From pressure law:

$\dfrac{P_1}{T_1} = \dfrac{P_2}{T_2}$

$\dfrac{2.5}{210} = \dfrac{P_2}{360}$

$210P_2 = 2.5 \times 360$

$P_2 = \dfrac{2.5 \times 360}{210}$

= 4.3

The pressure of the gas is 4.3atm

6. 500cm³ of air has a temperature of 141K at a pressure of 275mmHg. Calculate the temperature of the air when the pressure is increased to 495mmHg at the volume of 500cm³.

Solution

The volume of 500cm³ did not change (i.e. it was kept constant). So, this represents pressure law.

The given conditions are: $P_1 = 275$, $T_1 = 141$, $P_2 = 495$, $T_2 = ?$

From Pressure law:

$$\frac{P_1}{T_1} = \frac{P_2}{T_2}$$

$$\frac{275}{141} = \frac{495}{T_2}$$

$275T_2 = 141 \times 495$

$$T_2 = \frac{141 \times 495}{275}$$

$T_2 = 254$

The temperature of the air is 254K

7. A gas occupies a pressure of 0.95atm at a given temperature. Calculate the pressure of the gas when the temperature is halved at constant volume.

Solution

The given conditions are: $P_1 = 0.95$, Initial temperature = T_1 (Since the value is not given), $T_2 = T_1/2$ (Since the initial temperature is halved, i.e. divided by two to give the final temperature), $P_2 = ?$

From Pressure law:

$$\frac{P_1}{T_1} = \frac{P_2}{T_2}$$

$$\frac{0.95}{T_1} = \frac{P_2}{T_1/2}$$

Cross multiply to get:

$$0.95 \times \frac{T_1}{2} = P_2 \times T_1$$

$$\frac{0.95T_1}{2} = P_2 T_1$$

Dividing both sides by T_1 will cancel out T_1 to give:

$$\frac{0.95}{2} = P_2$$

$$P_2 = 0.475$$

The pressure of the gas is 0.0.475atm

Just like in Charles' law, the example above shows that in pressure law, when a quantity is increased or decreased by a factor, the other quantity will also be increased or decreased by the same factor. Study the examples below in order to understand this rule better.

8. The pressure of a fixed mass of gas is 210cmHg. Determine its pressure when its temperature is increased three times at constant volume.

Solution

When the temperature is increased three times, it means that the pressure will also increase three times.

∴ The new pressure is = 210 x 3

= 630cmHg

9. What is the new temperature (in °C) of a fixed mass of gas originally at 540K when its pressure is decreased nine times at the same volume?

61

Solution

When the pressure is decreased nine times, it means that the temperature will also decrease nine times.

∴ The new temperature in Kelvin is = $\dfrac{540}{9}$

= 60K

∴ The new temperature in °C = 60 – 273 = -213°C

10. The temperature of a given mass of air is 79°C. Determine its temperature in °C if its pressure is quartered at the same volume?

Solution

Temperature in Kelvin = 79 + 273 = 352K

When the pressure is quartered, it means that the temperature (in Kelvin) will also be quartered (i.e. divided by four).

∴ The new temperature is = $\dfrac{352}{4}$

= 88K

∴ The new temperature in °C = 88 – 273 = -185°C

Exercise 21

1. A fixed mass of gas occupies a pressure of 690mmHg at 80°C. Calculate its pressure when the temperature is increased to 142°C at constant volume.

2. A fixed mass of a gas occupies a pressure of 44cmHg at 110°C. Calculate its temperature when its pressure becomes 50cmHg.

3. Air at a temperature of 127°C occupies a pressure of 838mmHg. Calculate the temperature of the air in °C when its pressure is reduced to 400mmHg at constant volume.

4. A given mass of gas at a temperature of -86°C occupies a pressure of 700mmHg. Calculate the temperature of the gas in °C when its pressure is increased to 940mmHg at the same volume.

5. A fixed mass of gas occupies a pressure of 1.8atm at 390K. Calculate its pressure when the temperature is increased to 500K at constant volume.

6. 210cm^3 of air has a temperature of 192K at a pressure of 350mmHg. Calculate the temperature of the air when the pressure is increased to 650mmHg at the volume of 210cm^3.

7. A gas occupies a pressure of 2.15atm at a given temperature in Kelvin. Calculate the pressure of the gas when the temperature in Kelvin is halved at constant volume.

8. The pressure of a fixed mass of gas is 104cmHg. Determine its pressure when its temperature in Kelvin is increased five times at constant volume.

9. What is the new temperature (in °C) of a fixed mass of gas originally at 310K when its pressure is decreased ten times at the same volume?

10. The temperature of a given mass of air is 124°C. Determine its temperature in °C if its pressure is quartered at the same volume?

CHAPTER 22
GENERAL GAS LAW

When any two of Boyle's law, Charles' law and pressure law are combined together, they give the general gas law, which is expressed as:

$$\frac{PV}{T} = K$$

If the pressure, volume or temperature of a gas changes from an initial value to a final value, then,

$$\frac{P_1 V_1}{T_1} = \frac{P_2 V_2}{T_2}$$

Where 1 represent the initial values, and 2 represent the final values.

Note that s.t.p means standard temperature and pressure.
Standard temperature = 273K (i.e. 0°C)
Standard pressure = 1.01×10^5 Pa or 760mmHg or 76cmHg or 1atm.

Examples

1. A fixed mass of gas occupies a volume of 250cm³ at 27°C and 820mmHg. Calculate the volume of the gas at 100°C and 580mmHg.

Solution

Given from the question are: V_1 = 250, T_1 = 27 + 273 = 300, P_1 = 820, T_2 = 100 + 273 = 373, P_2 = 580, V_2 = ?

From the general gas law:

$$\frac{P_1 V_1}{T_1} = \frac{P_2 V_2}{T_2}$$

$$\frac{820 \times 250}{300} = \frac{580 \times V_2}{373}$$

Cross multiply to get:

$300 \times 580 \times V_2 = 820 \times 250 \times 373$

$174000 V_2 = 820 \times 250 \times 373$

$\therefore V_2 = \dfrac{820 \times 250 \times 373}{174000}$

$= 439$

The volume of the gas is 439cm^3

2. 125cm^3 of a given mass of gas is at a temperature of 47°C and a pressure of 120cmHg. Determine its temperature in °C if its volume becomes 200cm^3 at 30cmHg.

Solution

Given from the question are: $V_1 = 125$, $T_1 = 47 + 273 = 320$, $P_1 = 120$, $V_2 = 200$, $P_2 = 30$, $T_2 = ?$

From the general gas law:

$$\dfrac{P_1 V_1}{T_1} = \dfrac{P_2 V_2}{T_2}$$

$$\dfrac{120 \times 125}{320} = \dfrac{30 \times 200}{T_2}$$

Cross multiply to get:

$120 \times 125 \times T_2 = 30 \times 200 \times 320$

$15000 T_2 = 30 \times 200 \times 320$

$\therefore T_2 = \dfrac{30 \times 200 \times 320}{15000}$

$= 128K$

The temperature of the gas in °C = 128 - 273 = -145°C

3. A fixed mass of air occupies a volume of 315cm³ at 650mmHg and at an unknown temperature. If the air is converted to occupy a volume of 280cm³ at 800mmHg and a temperature of 210°C, determine the unknown temperature.

Solution

Given from the question are: V_1 = 315, P_1 = 650, T_1 = ?, V_2 = 280, P_2 = 800, T_2 = 210 + 273 = 483.

From the general gas law:

$$\frac{P_1 V_1}{T_1} = \frac{P_2 V_2}{T_2}$$

$$\frac{650 \times 315}{T_1} = \frac{800 \times 280}{483}$$

Cross multiply to get:

800 x 280 x T_1 = 650 x 315 x 483

224000T_1 = 650 x 315 x 483

∴ $T_1 = \dfrac{650 \times 315 \times 483}{224000}$

= 441K

4. The volume of a fixed mass of gas is 57cm³ at 27°C and 80cmHg. What is its volume at s.t.p?

Solution

Given from the question are: V_1 = 57, T_1 = 27 + 273 = 300, P_1 = 80, T_2 = s.t = 273K, P_2 = s.p = 76cmHg, V_2 = ?

Note that s.t means standard temperature and s.p means standard pressure. The value of the standard pressure chosen must be in the same unit as that given in the question. In this case it is pressure in cmHg.

$$\therefore \quad \frac{P_1 V_1}{T_1} = \frac{P_2 V_2}{T_2}$$

$$\frac{80 \times 57}{300} = \frac{76 \times V_2}{273}$$

Cross multiply to get:

300 x 76 x V₂ = 80 x 57 x 273

22800V₂ = 80 x 57 x 273

$$\therefore \quad V_2 = \frac{80 \times 57 \times 273}{22800}$$

= 54.6

The volume of the gas is 54.6cm³

5. At s.t.p the volume of a gas is 300cm³. Calculate its temperature if its volume changes to 450cm³ at a pressure of 650mmHg.

Solution

T_1 = S.t = 273K, P_1 = s.p = 760mmHg, V_1 = 300, V_2 = 450, P_2 = 650, T_2 = ?

Note that the value of the standard pressure chosen is in the same unit as that given in the question. In mmHg, standard pressure is 760mmHg.

$$\therefore \quad \frac{P_1 V_1}{T_1} = \frac{P_2 V_2}{T_2}$$

$$\frac{760 \times 300}{273} = \frac{650 \times 450}{T_2}$$

Cross multiply to get:

$$760 \times 300 \times T_2 = 273 \times 650 \times 450$$

$$228000 T_2 = 273 \times 650 \times 450$$

$$\therefore \quad T_2 = \frac{273 \times 650 \times 450}{228000}$$

$$= 350K$$

The temperature of the gas in °C = 350 - 273 = 77°C

6. Air at a temperature of 527°C and 30atm is admitted into the cylinder of an engine. Calculate the pressure of the air when it has expanded to five times its volume and cooled to 127°C as it leaves the engine.

Solution

Initial volume = V_1 (Since the value is not stated), T_1 = 527 + 273 = 800, P_1 = 30, T_2 = 127 + 273 = 400, $V_2 = 5V_1$, (Since the new volume is five times the initial volume), P_2 = ?

$$\therefore \quad \frac{P_1 V_1}{T_1} = \frac{P_2 V_2}{T_2}$$

$$\frac{30 \times V_1}{800} = \frac{P_2 \times 5V_1}{400}$$

Cross multiply to get:

$$400 \times 30 \times V_1 = 800 \times P_2 \times 5V_1$$

Dividing both sides by V_1 will cancel out V_1 to give:

$$400 \times 30 = 800 \times P_2 \times 5$$

$$12000 = 4000 P_2$$

$$\therefore \quad P_2 = \frac{12000}{4000}$$

$$P_2 = 3$$

The pressure of the air is 3atm.

7. A fixed mass of gas is at a temperature of 200K. Calculate the temperature of the gas in °C if its original volume is reduced five times while its pressure is doubled.

Solution

Initial volume = V_1 (Since the value is not stated), T_1 = 200, Initial pressure = P_1 (Since the value is not stated), $V_2 = V_1/5$ (Since the original volume is reduced five times), P_2 = $2P_1$, (Since the new pressure is 2 times the initial pressure, i.e. doubled), T_2 = ?

∴ $$\frac{P_1V_1}{T_1} = \frac{P_2V_2}{T_2}$$

$$\frac{P_1 \times V_1}{200} = \frac{2P_1 \times V_1/5}{T_2}$$

Cross multiply to get:

$$T_2 \times P_1 \times V_1 = 200 \times 2P_1 \times \frac{V_1}{5}$$

Dividing both sides by P_1V_1 will cancel out P_1V_1 to give:

$$T_2 = \frac{200 \times 2}{5}$$

T_2 = 80K

The temperature of the gas in °C = 80 -273 = -193°C

8. The pressure of the molecules of a gas in an adjustable container is 805mmHg. The volume of the container is adjusted to thrice its initial value while the temperature (in Kelvin) of the gas molecules is doubled. Determine the new pressure of the gas molecules.

Solutions

Initial volume = V_1, P_1 = 805, Initial temperature = T_1, $V_2 = 3V_1$, $T_2 = 2T_1$, P_2 = ?

∴ $$\frac{P_1V_1}{T_1} = \frac{P_2V_2}{T_2}$$

$$\frac{805 \times V_1}{T_1} = \frac{P_2 \times 3V_1}{2T_1}$$

Cross multiply to get:

805 x V₁ x 2T₁ = T₁ x P₂ x 3V₁

Dividing both sides by V₁T₁ will cancel out V₁T₁ to give:

805 x 2 = 3P₂

$$P_2 = \frac{805 \times 2}{3}$$

P₂ = 537

The new pressure of the gas molecules is 537mmHg.

Exercise 22

1. A fixed mass of gas occupies a volume of 100cm³ at 57°C and 610mmHg. Calculate the volume of the gas at 80°C and 805mmHg.

2. 200cm³ of a given mass of gas is at a temperature of 62°C and a pressure of 92cmHg. Determine its temperature in °C if its volume becomes 250cm³ at 160cmHg.

3. A fixed mass of air occupies a volume of 65cm³ at 900mmHg and at an unknown temperature. If the air is converted to occupy a volume of 140cm³ at 850mmHg and a temperature of 155°C, determine the unknown temperature.

4. The volume of a fixed mass of gas is 200cm³ at 27°C and 71cmHg. What is its volume at s.t.p?

5. At s.t.p, the volume of a gas is 90cm³. Calculate its temperature if its volume increases to 340cm³ at a pressure of 480mmHg.

6. Air at a temperature of 407°C and 18atm is admitted into the cylinder of an engine. Calculate the pressure of the air when it has expanded to two times its volume and cooled to 172°C as it leaves the engine.

7. A fixed mass of gas is at a temperature of 320K. Calculate the temperature of the gas in °C if its original volume is reduced three times while its pressure is quadrupled.

8. The pressure of the molecules of a gas in an adjustable container is 70cmHg. The volume of the container is adjusted to five times its initial value while the temperature (in Kelvin) of the gas molecules is doubled. Determine the new pressure of the gas molecules.

ANSWERS TO EXERCISES

Exercise 1
(1)(a) 3.14sec (b) 0.318Hz (c) 0.2m/s (d) 0.4m/s^2 (e) 0.173m/s
(2)(a) 1.99sec (b) 0.316m/s (3)(a) 0.72m/s (b) 1.02m/s (4) 0.71Hz
(5) 6sec (6)(a) 0.02sec (b) 18.85m/s(middle). The velocity at the end is zero
(7)(a) 2.97m (b) 14.96m/s^2 (8) 28.3sec (9) 7.2sec (10) 4.08rad/sec
(11)(a) 10rad/sec (b) 0.98m/s
(12)(a) A = 10m, f = 1Hz, T = 1sec (b) $v = -62.84\sin(2\pi t + \frac{\pi}{4})$, $a = -394.9\cos(2\pi t + \frac{\pi}{4})$
(c) x = 7.07m, v = −44.4m/s, a = −279.2m/s^2
(13)(a) 32m/s (b) 128m/s^2 (c) 10.2m (14) 6sec (15) 1 : 4

Exercise 2
(1)(a) 0.281sec (b) 0.05J (c) 0.05J (2)(a) 0.397sec (b) 0.0225J
(3)(a) K.E = 3.2 x 10^{-6}J, P.E = = 4 x 10^{-7}J (b) = 3.6 x 10^{-6}J
(4)(a) 8.1cm (b) P.E = 0.00197J, K.E = 0.0032J (c) 0.00517J
(5) 0.00216J (6) 0.000222J (7)(a) K.E = 2.49 x 10^{-6}J, P.E = 4.74 x 10^{-7}J
(b) 2.96 x 10^{-6}J (8)(a) 0.562sec (b) 0.002J (c) 0.002J

Exercise 3
(1) 2m/s^2 (2) 40Hz (3) 0.21rad/sec (4)(a) 0.167sec (b) 37.6rad/sec
(c) 7.5m/s (d) 282.8m/s^2 (e) 1414s^{-2} (5)(a) 0.786rad/sec
(b) 0.628m/s (c) 0.125Hz (d) 0.492m/s^2 (6) 12.38N (7) 0.123N
(8)(a) 0.97rad/sec (b) 0.116m/s (c) 0.00113N

Exercise 4
(1)(a) A = 200cm (or 2m), λ = 160cm (or 1.6m) (b) 1.26sec (c) 1.27m/s
(2) 3.2m (3) 0.375m (4) 0.46Hz (5) 0.0533sec
(6)(a) 10m (b) 4Hz (c) 50m/s (d) 12.5m (e) 0.503m^{-1}
(7)(a) 20m (b) 0.01sec (c) 500m/s (d) $y = 8\sin 2\pi(25t + \frac{x}{20})$
(8)(a) 4m (b) 0.318Hz (c) 3.14sec (d) 0.40m/s (e) 2.0rad/sec
(f) 1.26m (g) 5.0m^{-1} (9)(a) 0.07m (b) 0.2Hz (c) 0.24m/s
(10)(a) 1m (b) 1.25m/s (c) $y = 1.5\sin 2\pi(1.25t + x)$

Exercise 5
(1) 660m   (2) 7.06sec   (3) 332m/s   (4)(a) 1.71sec   (b) 2.35sec
(5) 49   (6) 324m/s   (7) 338m/s   (8) 378m/s

Exercise 6
(1)(a) 4Hz   (b) 0.25sec   (2)(a) 1.25Hz   (b) 53.25Hz   (3) 6Hz
(4) 23Hz and 17Hz   (5)(a) 0.71Hz   (b) 324.3Hz

Exercise 7
(1) 196.4Hz   (2) λ = 45cm, f = 1511Hz   (3) λ = 0.2m, f = 1020Hz
(4) 1320Hz   (5) 8.05m   (6) 364m/s   (7) 333.8m/s   (8) 354.2Hz

Exercise 8
(1) 68.5Hz   (2) 100.6Hz   (3) 8.5N   (4) 111.8Hz   (5) 297Hz
(6) 62Hz   (7)(a) 41.4Hz   (b) 69Hz   (c) 58m/s   (8) 55Hz
(9) 14.4Hz   (10) 50cm

Exercise 9
(1) 135Hz   (2) 0.88rec/sec   (3) 37 teeth   (4) 22.2sec   (5) 152sec

Exercise 10
(1) 232.3Hz   (2) 357.3Hz   (3) 450.2Hz   (4) 375Hz   (5) 460Hz
(6) 347Hz   (7) 398.4Hz   (8) 441.4Hz   (9) 300.7m/s   (10) 3.6m/s

Exercise 11
(1) 8.436 x 10^{-3}m   (2) 1.894 x 10^{-5}/k   (3) 219.9cm   (4) 136.7°C
(5) 1515.2°C   (6) 7.2 x 10^{-4}m   (7) 3 : 2

Exercise 12
(1) 0.15cm^2 (2) 2.36 x 10^{-5}/k   (3) 1197.9cm^2 (4) 54.4°C
(5) 3.62 x 10^{-5}/k   (6) 52.5cm^2 (7) 51.625°C

Exercise 13
(1) 2.67 x 10^{-5}/k   (2) 0.05244m^3 (3) 6.54 x 10^{-5}/k   (4) 1942.4mm^3
(5) 43.94°C   (6) 1.82 x 10^{-5}/k   (7) 3,570mm^3

Exercise 14
1(a) 5.39×10^{-4}/k (b) 5.61×10^{-4}/k 2(a) $8.7 \times 10^{-4} cm^3$ (b) 6.21cm
(3) 1.96×10^{-4}/k (4) 304.49g 5(a) 166.4g (b) 12.7g
(6) 1.56×10^{-4}/k (7) 3.06×10^{-4}/k

Exercise 15
(1) 40.65°C 2(a) 16.7°C (b) 104 3(a) 55.9°C (b) 14.65Ω
(4) 43.75°C 5(a) 113°F (b) 15°C 6(a) 38.1°C (b) 278mm
7(a) 66.7°C (b) 17.5cm 8(a) 62.5°C (b) 12Ω

Exercise 16
(1) 5850J (2) 93°C (3) 66°C (4) 7.6kg (5) 40.8°C (6) 34.1°C
7(a) 13J/s (b) $722.2 Jkg^{-1}k^{-1}$ (8) 672sec (9) 2.75kg (10) 168.75°C
(11) 68.9sec (12) 74°C

Exercise 17
(1) 16,800J (2) 6517.5J (3) 313.4g (4) 246600J/kg (5) 573,942.5J
(6) 5.74A (7) 40.1V (8) 73.8°C (9) 89°C (10) 8.2sec

Exercise 18
(1) 66.7% (2) 6.15g (3) 52.3% (4) 2.87cmHg (5) 2.173g
(6) 2.42cmHg

Exercise 19
1(a) 22.74cm (b) 30.86cm (2) 70.9cm 3(a) 83.3cmHg (b) 7.38cm
(4) $313.6 cm^3$ (5) 171.1cmHg (6) 200mmHg (7) 6800mm3
(8) $124 cm^3$ (9) 450mmHg (10) The pressure will be increased twenty times
(11) 162mmHg (12) $172.7 cm^3$

Exercise 20
(1) $562.5 cm^3$ (2) 705.1K or 432.1°C (3) 303.9°C (4) 238.2°C
(5) $53.3 cm^3$ (6) 1400K (7) $0.525 dm^3$ (8) $750 cm^3$ (9) −118°C
(10) 1257°C

Exercise 21
(1) 811.2mmHg (2) 435.2K or 162.2°C (3) − 82.1°C (4) −21.8°C

(5) 2.3atm (6) 356.6K (7) 1.075atm (8) 520cmHg
(9) −242°C (10) −173.75°C

Exercise 22
(1) 81.06cm^3 (2) 455.3°C (3) −62.6°C (4) 170.02cm^3
(5) 651.4K or 378.4°C (6) 11.8atm (7) 153.7°C (8) 28cmHg

If you have any enquiries, suggestions, corrections or information concerning this book, please contact the author through the email below.

KINGSLEY AUGUSTINE

kingzohb2@yahoo.com

Twitter handle: @kingzohb2

www.ingramcontent.com/pod-product-compliance
Lightning Source LLC
Chambersburg PA
CBHW080454220526
45465CB00006B/2270